Nitration and aromatic reactivity

Nitration and aromatic reactivity

J. G. HOGGETT, R. B. MOODIE,
J. R. PENTON & K. SCHOFIELD

Department of Chemistry
University of Exeter

CAMBRIDGE

AT THE UNIVERSITY PRESS · 1971

CAMBRIDGE UNIVERSITY PRESS
Cambridge, New York, Melbourne, Madrid, Cape Town, Singapore, São Paulo, Delhi

Cambridge University Press
The Edinburgh Building, Cambridge CB2 8RU, UK

Published in the United States of America by Cambridge University Press, New York

www.cambridge.org
Information on this title: www.cambridge.org/9780521104944

First published 1971
This digitally printed version 2009

A catalogue record for this publication is available from the British Library

Library of Congress Catalogue Card Number: 76–138374

ISBN 978-0-521-08029-3 hardback
ISBN 978-0-521-10494-4 paperback

Contents

Contents

Preface

Ten years ago we became interested in the possibility of using nitration as a process with which to study the reactivity of hetero-aromatic compounds towards electrophilic substitution. The choice of nitration was determined by the consideration that its mechanism was probably better understood than that of any other electrophilic substitution. Others also were pursuing the same objective, and a considerable amount of information has now been compiled.

This work, and that which has been reported in the past decade about the general topic of nitration, has advanced our knowledge appreciably, and has also revealed some gaps in our understanding of the subject. This book reviews the present position, and collects together much detailed information about quantitative aspects of nitration from the extensive research literature on the subject. In offering it we should like to express our gratitude to Drs R. G. Coombes, J. T. Gleghorn, S. R. Hartshorn, E. A. Qureshi, M. J. Thompson and M. J. Williamson, who, as well as our co-authors, have worked with us on nitration.

R. B. MOODIE
K. SCHOFIELD

1 Introduction

1.1 THE IMPORTANCE OF NITRATION

Nitration is important for two reasons: firstly, because it is the most general process for the preparation of aromatic nitro compounds; secondly, because of the part which it has played in the development of theoretical organic chemistry. It is of interest because of its own characteristics as an electrophilic substitution.

The first nitration to be reported was that of benzene itself. Mitscherlich in 1834 prepared nitrobenzene by treating benzene with fuming nitric acid.[1] Not long afterwards the important method of effecting nitration with a mixture of nitric and sulphuric acids ('mixed acid') was introduced, evidently in a patent by Mansfield;[2] the poor quality of early nitric acid was probably the reason why the method was developed. Since these beginnings, nitration has been the subject of continuous study.

1.2 NITRATING AGENTS

The means which have been used for effecting nitration are numerous,[3] but not all of the methods are in common use. Dilute nitric acid is useful for nitrating reactive substances such as phenol, but the oxidising properties of more concentrated nitric acid can be disadvantageous. Solutions of nitric acid or nitrates in sulphuric acid of various concentrations, or in oleum, provide reagents of a wide range of vigour. They have the additional property, often useful, that some organic compounds are appreciably soluble in them, and the disadvantage of being able to sulphonate some aromatic compounds. The disadvantage is rarely serious, for nitration is generally a more rapid process than sulphonation. Nitric acid in organic solvents also provides reagents in which aromatic compounds are usefully soluble, but these solutions are milder nitrating agents than those in mineral acids. In preparative nitration, acetic acid is probably the most frequently used of organic solvents. Solutions of nitric acid in organic solvents are less acidic than solutions in mineral acids, a virtue when compounds sensitive to acids are being nitrated, and one which is shared by solutions of nitric acid in acetic anhydride (these reactants react together fairly rapidly to give acetyl nitrate;

Introduction

§ 5.3). Even less dangerous in this respect are the nitrating systems using alkyl nitrates and sodium ethoxide. Noteworthy examples of the use of these less acidic or basic nitrating systems are found in the pyrrole series.[4a]

Nitronium salts in solution in inert organic solvents have been used in recent years to nitrate a wide range of aromatic compounds. Yields are generally good, but in preparative work the method is advantageous only in special cases, notably where the aromatic contains a hydrolysable substituent (§ 4.4).

In recent years the analogy between the Friedel–Crafts acylation reaction and various nitrating systems, particularly those in which Lewis acids act as catalysts, has been stressed,[3c] but this classification adds nothing new in principle.

Our special concern is with those nitrating systems for which mechanistic studies have established, or made probable, the identity of the electrophile through which they effect nitration. In most cases, though not quite all, this has proved to be the nitronium ion. The important nitrating systems formed by mixing nitric acid with acetic anhydride stand out in having so far resisted attempts to identify beyond reasonable doubt the electrophile or electrophiles through which they operate (§ 5.3).

These systems nitrate aromatic compounds by a process of electrophilic substitution, the character of which is now understood in some detail (§ 6.1). It should be noted, however, that some of them can cause nitration and various other reactions by less well understood processes. Among such nitrations that of nitration *via* nitrosation is especially important when the aromatic substrate is a reactive one (§4.3). In reaction with lithium nitrate in acetic anhydride, or with fuming nitric acid, quinoline gives a small yield of 3-nitroquinoline; this untypical orientation (cf. § 10.4.2[4b]) may be a consequence of nitration following nucleophilic addition.[5]

As regards reactions other than nitration brought about by some of these systems, especially noteworthy are the addition processes undergone by certain indole derivatives when treated with solutions of nitric acid in acetic acid. Products include glycols, nitro-alcohols, and nitro-alcohol acetates.[4b] Such additions might well be encountered with some polynuclear aromatic compounds, and with such compounds the possibility of nitration by addition–elimination must always be borne in mind.

2

Benzene and some of its derivatives react with solutions of mercuric nitrate in concentrated nitric acid to give nitrophenols. These reactions, known as oxynitrations may proceed by mercuration followed by nitroso-demercuration; the resulting nitroso compound becomes a diazonium compound and then a phenol, which is nitrated.[6]

1.3 NITRATION AND THE DEVELOPMENT OF THEORETICAL ORGANIC CHEMISTRY

The development of theoretical organic chemistry was intimately entwined with the development of that particular aspect of it concerned with aromatic substitution; the history of this twin growth has been authoritatively traced.[7] Only the main developments, particularly as they affect nitration, will be noted here.

With the establishment of structural organic chemistry in the nineteenth century, the first major landmark in its development to include reactivity was the publication by Holleman in 1910 of his studies on orientation in aromatic substitution.[8a] The reactions considered by Holleman were those which we now recognise as electrophilic substitutions, and by far the most extensive data were those relating to nitration. Thus, the reaction played a major role in the recognition of the main generalisations which can be made about orientation: the classification of orienting substituents as *o:p*- or *m*-directing (already recognised by earlier workers); the frequent association of *o*- with dominant *m*-substitution; the product rule describing further substitution into di-substituted benzenes.[8] Of the greatest importance was Holleman's recognition of the connection between orientation and activation; *o : p*-orienting substituents commonly activate the aromatic nucleus, whilst *m*-orienting substituents de-activate it.

It is interesting to recognise why nitration was so pre-eminently useful a reaction for these purposes. First, it is very generally applicable because of the variety of conditions under which it can be carried out; second, it can usually be stopped cleanly after the first stage, because the nitro group introduced in the first stage is so powerfully deactivating; and third, because despite the wide range of conditions which may be used, nitration most commonly proceeds through the agency of the nitronium ion.

For the electronic theory of organic chemistry 1926 was the *annus mirabilis*, and, particularly, as they applied to aromatic substitution, the

Introduction

ideas of Lapworth, Robinson, and Ingold[9] approached their definitive form,[10] mainly through studies of nitration. Especially important in the refinement and critical testing of these ideas was the combination of data on orientation with data on rate; the definition of partial rate factors, originally called coefficients of activation,[11] required the theory to give an account of the state of activation of each individual position in an aromatic molecule.[11] As will be seen, nitration continues to be a testing ground of theoretical ideas, as in clarifying the nature of the inductive effect (§9.1.2), and in further defining the notion of aromatic reactivity (§7.1.2).

The electronic theory of organic chemistry, and other developments such as resonance theory, and parallel developments in molecular orbital theory relating to aromatic reactivity have been described frequently. A general discussion here would be superfluous; at the appropriate point a brief summary of the ideas used in this book will be given (§7.11).

REFERENCES

1. Mitscherlich, E. (1834). *Annln Phys. Chem.* **31**, 625; *Annln Pharm.* **12**, 305.
2. Roscoe, H. E. & Schorlemmer, C. (1891). *A Treatise on Chemistry*, vol. 3, p. 102.
3. (a) J. Houben (ed.) (1924). *Die Methoden der Organischen Chemie (Weyls Methoden)*, vol. 4, pp. 102 ff. Leipzig.
    (b) Topchiev, A. V. (1959). *Nitration of Hydrocarbons and Other Organic Compounds*. London: Pergamon Press.
    (c) Olah, G. A. & Kuhn, S. J. (1964). In *Friedel–Crafts and Related Reactions* (ed. G. A. Olah), ch. 43. New York: Interscience.
4. (a) Schofield, K. (1967). *Hetero-Aromatic Nitrogen Compounds: Pyrroles and Pyridines*, pp. 79–80. London: Butterworth.
    (b) Schofield, K. (1950). *Q. Rev. chem. Soc.* **4**, 382.
5. Dewar, M. J. S. & Maitlis, P. M. (1955). *Chemy Ind.* p. 685; (1957). *J. chem. Soc.* p. 944.
6. Wolffenstein, R. & Böters, O. (1913). *Ber. dt. chem. Ges.* **46**, 586.
    Westheimer, F. H., Segel, E. & Schramm, R. (1947). *J. Am. chem. Soc.* **69**, 773.
    Carmack, M., Baizer, N. M., Handrick, G. R., Kissinger, L. W. & Specht, E. H. (1947). *J. Am. chem. Soc.* **69**, 785.
    Wolffenstein, R. & Paar, W. (1913). *Ber. dt. chem. Ges.* **46**, 589.
7. Ingold, C. K. (1953). *Structure and Mechanism in Organic Chemistry*, ch. 6. London: Bell.
    Ingold, C. K. (1954). *Chemistry of Carbon Compounds* (ed. E. H. Rodd), vol. 3A, ch. 1. London: Elsevier.

8. (*a*) Holleman, A. F. (1910). *Die direkte Einführung von Substituenten in den Benzolkern.* Leipzig: Veit.
   (*b*) (1925). *Chem. Rev.* **1**, 187.
9. Robinson, R. (1932). *Two Lectures on an Outline of an Electrochemical (Electronic) Theory of the Course of Organic Reactions.* London: Institute of Chemistry.
   Ingold, C. K. (1934). *Chem. Rev.* **15**, 225.
10. Allan, J. & Robinson, R. (1926). *J. chem. Soc.* p. 376.
    Oxford, A. E. & Robinson, R. (1926). *J. chem. Soc.* p. 383.
    Robinson, R. & Smith, J. C. (1926). *J. chem. Soc.* p. 392.
    Allan, J., Oxford, A. E., Robinson, R. & Smith, J. C. (1926). *J. chem. Soc.* p. 401.
    Lea, T. R. & Robinson, R. (1926). *J. chem. Soc.* p. 411.
    Ing, H. R. & Robinson, R. (1926). *J. chem. Soc.* p. 1655.
    Ingold, C. K. & Ingold, E. H. (1926). *J. chem. Soc.* p. 1310.
    Goss, F. R., Ingold, C. K. & Wilson, I. S. (1926). *J. chem. Soc.* p. 2440.
    Ingold, C. K. (1926). *A. Rep. chem. Soc.* **23**, 129.
11. Ingold, C. K. & Shaw, F. R. (1927). *J. chem. Soc.* p. 2918.

2 Nitrating systems:
A. Mineral acids

2.1 INTRODUCTION

Nitration can be effected under a wide variety of conditions, as already indicated. The characteristics and kinetics exhibited by the reactions depend on the reagents used, but, as the mechanisms have been elucidated, the surprising fact has emerged that the nitronium ion is pre-eminently effective as the electrophilic species. The evidence for the operation of other electrophiles will be discussed, but it can be said that the supremacy of one electrophile is uncharacteristic of electrophilic substitutions, and bestows on nitration great utility as a model reaction.

Euler[1] first suggested that the nitronium ion was the active species, but proof of this did not come for many years. Investigations of the mechanisms have been chiefly concerned with the physical examination of the media used, and with the kinetics of the reactions.

2.2 NITRATION IN CONCENTRATED AND AQUEOUS NITRIC ACID

2.2.1 *The state of concentrated nitric acid*

Molecular nitric acid is the main species present in this medium, but physical measurements demonstrate the existence of significant concentrations of other species.

By studying the variation of the freezing point of mixtures of di-nitrogen pentoxide and water, over a range of concentration encompassing the formation of pure nitric acid, it was shown[2] that appreciable self-dehydration was occurring according to the following scheme:

$$2HNO_3 \rightleftharpoons NO_2^+ + NO_3^- + H_2O.$$

At the freezing point of nitric acid ($-42\ °C$), the concentrations of water, nitronium ion, and nitrate ion were found[2] to be 0.41 mol l^{-1}, but more recent work[3] suggests the value 0.69 mol l^{-1}. Measurements of the electrical conductivity of nitric acid at $-10\ °C$ give a value of 0.51 mol l^{-1}, and at $-20\ °C$ of 0.61 mol l^{-1} for the concentrations of the three species.[4]

6

The Raman spectrum of nitric acid shows[5] two weak bands at 1050 and 1400 cm^{-1}. By comparison with the spectra of isolated nitronium salts[6] (§2.3.1), these bonds were attributed to the nitrate and nitronium ion respectively. Solutions of dinitrogen pentoxide in nitric acid show these bands[7], but not those characteristic of the covalent anhydride[8], indicating that the self-dehydration of nitric acid does not lead to molecular dinitrogen pentoxide. Later work on the Raman spectrum indicates that at -15 °C the concentrations of nitrate and nitronium ion are 0·37 mol l^{-1} and 0·34 mol l^{-1}, respectively.[9] The infra-red spectrum of nitric acid shows absorption bands characteristic of the nitronium ion.[10] The equivalence of the concentrations of nitronium and nitrate ions argues against the importance of the following equilibrium:

$$2HNO_3 \rightleftharpoons H_2NO_3^+ + NO_3^-.$$

2.2.2. *The state of aqueous solutions of nitric acid*

In strongly acidic solutions water is a weaker base than its behaviour in dilute solutions would predict, for it is almost unprotonated in concentrated nitric acid,[9] and only partially protonated in concentrated sulphuric acid.[11a] The addition of water to nitric acid affects the equilibrium leading to the formation of the nitronium and nitrate ions (§2.2.1). The intensity of the peak in the Raman spectrum associated with the nitronium ion decreases with the progressive addition of water, and the peak is absent from the spectrum of solutions containing more than about 5 % of water;[5a] a similar effect has been observed in the infra-red spectrum.[10]

Because water is not protonated in these solutions, its addition reduces the concentration of ions, and therefore the electrical conductivity. The conductivity reaches a minimum in solutions containing 97 % of acid, but rises on further dilution as a result of the formation of nitrate and hydroxonium ions.[4]

The infra-red absorption bands of molecular nitric acid do not change as the medium is varied beween 100 % and 70 % of acid; on further dilution the nitrate ion becomes the dominant species.[10]

In equimolar mixtures of nitric acid and water a monohydrate is formed whose Raman spectrum has been observed.[12] There is no evidence for the existence of appreciable concentrations of the nitric acidium ion in aqueous nitric acid.

2.2.3 *Nitration in concentrated nitric acid*

Our knowledge of the mechanism of the reaction in this medium comes from an investigation of the nitration of nitrobenzene, *p*-chloronitrobenzene and 1-nitroanthraquinone.[13] These compounds underwent reaction according to the following rate law:

$$\text{rate} = k_1[\text{ArH}]$$

Nitric acid being the solvent, terms involving its concentration cannot enter the rate equation. This form of the rate equation is consistent with reaction *via* molecular nitric acid, or any species whose concentration throughout the reaction bears a constant ratio to the stoichiometric concentration of nitric acid. In the latter case the nitrating agent may account for any fraction of the total concentration of acid, provided that it is formed quickly relative to the speed of nitration. More detailed information about the mechanism was obtained from the effects of certain added species on the rate of reaction.

Sulphuric acid catalysed nitration in concentrated nitric acid, but the effect was much weaker than that observed in nitration in organic solvents (§ 3.2.3). The concentration of sulphuric acid required to double the rate of nitration of 1-nitroanthraquinone was about 0·23 mol l^{-1}, whereas typically, a concentration of 10^{-3} mol l^{-1} will effect the same change in nitration in mixtures of nitric acid and organic solvents. The acceleration in the rate was not linear in the concentration of catalyst, for the sensitivity to catalysis was small with low concentrations of sulphuric acid, but increased with the progressive addition of more catalyst and eventually approached a linear acceleration.

Potassium nitrate anticatalysed nitration in nitric acid (the solutions used also contained 2·5 mol l^{-1} of water) but the effect was small in comparison with the corresponding effect in nitration in organic solvents (§ 3.2.3 & 4), for the rate was only halved by the addition of 0·31 mol l^{-1} of the salt. As in the case of the addition of sulphuric acid, the effect was not linear in the concentration of the additive, and the variation of $k^{-1}/$s with [KNO$_3$]/mol l^{-1} was similar to that of $k_1/$s^{-1} with [H$_2$SO$_4$]/mol l^{-1}.

The relative weakness of the two effects, and the adoption of the kinetic form of the catalysis to the linear law only when the concentration of the additive was greater than *c*. 0·2 mol l^{-1}, results from the equilibria existing in anhydrous nitric acid. In the absence of catalyst, nitric acid undergoes appreciable self-dehydration to yield nitronium ions, nitrate ions and water. The addition of sulphuric acid allows the operation of another mode of ionization:

$$HNO_3 + H_2SO_4 \rightleftharpoons NO_2^+ + HSO_4^- + H_2O.$$

The nitronium ions produced in this way tend to repress the self-dehydration of the nitric acid and therefore the net concentration of nitronium ions is not proportional to the concentration of the catalyst. When sufficient sulphuric acid has been added to make the self-ionization of nitric acid relatively unimportant, the nitronium ions will be produced predominantly from the above ionization, and the acceleration will follow a linear law.

The effect of potassium nitrate on the rate arises in a similar way. The concentration of nitrate ions in concentrated nitric acid is appreciable, and addition of small quantities of nitrate will have relatively little effect. Only when the concentration of added nitrate exceeds that of the nitrate present in pure nitric acid will the anticatalysis become proportional to the concentration of added salt.

Therefore, in the cases of both additives, the kinetic law for the catalysis will assume a linear form when the concentration of the added species, or, in the case of sulphuric acid, the nitronium ion generated by its action, is comparable with the concentration of the species already present. This effect was observed to occur when the concentration of additive was about 0.2 mol l^{-1}, a value in fair agreement with the estimated degree of dissociation of nitric acid (§ 2.2.1).

2.2.4 *Nitration in aqueous solutions of nitric acid*

Added water retards nitration in concentrated nitric acid without disturbing the kinetic order of the reaction.[13] The rate of nitration of nitrobenzene was depressed sixfold by the addition of 5% of water, ($c.$ 3.2 mol l^{-1}), but because of the complexity of the equilibria involving water, which exist in these media, no simple relationship could be found between the concentration of water and its effect on the rate.

In more dilute solutions the concentration of the nitronium ion falls below the level of spectroscopic detection, and the nature of the electrophilic species has been the subject of conjecture.

The nitration of 2-phenylethanesulphonate anion (I) and toluene-ω-sulphonate anion (II) in aqueous nitric acid containing some added perchloric or sulphuric acid has been studied.[14] When the medium was varied the rate of reaction did not change according to the concentration of molecular nitric acid, which could not therefore be the active species. The distinction between the operation of the nitronium ion and the nitric acidium ion $H_2NO_3^+$ was less easy to make. The authors of the above work preferred the claims of the latter but could not exclude reaction *via* the small concentration of nitronium ions.

(I)   (II)   (III)   (IV)

(V)

The operation of the nitronium ion in these media was later proved conclusively.[15a-c] The rates of nitration of 2-phenylethanesulphonate anion ([Aromatic] < *c.* 0·5 mol l^{-1}), toluene-ω-sulphonate anion, *p*-nitrophenol, *N*-methyl-2,4-dinitroaniline and *N*-methyl-*N*,2,4-trinitro-aniline in aqueous solutions of nitric acid depend on the first power of the concentration of the aromatic.[15b] The dependence on acidity of the rate of ^{18}O-exchange between nitric acid and water was measured,[15a] and formal first-order rate constants for oxygen exchange were defined by dividing the rates of exchange by the concentration of water.[15b] Comparison of these constants with the corresponding results for the reactions of the aromatic compounds yielded the scale of relative reactivities shown in table 2.1.

When the concentration of 2-phenylethanesulphonate anion was > 0·5 mol l^{-1}, or when 2-mesitylethanesulphonate anion (v),[15c] mesitylene-α-sulphonate anion, or iso-durene-α^2-sulphonate anion[15b] were nitrated, the initial part of the reaction deviated from a first-order dependence on the concentration of the aromatic towards a zeroth-order dependence.

TABLE 2.1 *Relative rates of nitration in aqueous nitric acid**

Compound	Relative rate
iso-Durene-α^2-sulphonate anion (IV)	∼ 2200
Mesitylene-α-sulphonate anion (III)	∼ 1500
2-Phenylethanesulphonate anion (I)	100–250
N-Methyl-2,4-dinitroaniline	100–250
Toluene-ω-sulphonate anion (II)	20–45
p-Nitrophenol	10–13
Water	1
N-Methyl-N,2,4-trinitroaniline	0·04

* The values are 'precisely defined only for a given nitric acid concentration, as the dependence of rate on nitric acid concentration varies from one compound to another'.[15b]

For the last two compounds, first-order rates were observed towards the end of the reactions, enabling the reactivities of these compounds relative to that of water to be estimated (table 2.1). The nitration of 2-mesitylethanesulphonate anion was independent of the concentration of the aromatic over 80 % of its course, and because the final part of the reaction did not obey a truly first-order law its reactivity could not be estimated.

Nitration at a rate independent of the concentration of the compound being nitrated had previously been observed in reactions in organic solvents (§3.2.1). Such kinetics would be observed if the bulk reactivity of the aromatic towards the nitrating species exceeded that of water, and the measured rate would then be the rate of production of the nitrating species. The identification of the slow reaction with the formation of the nitronium ion followed from the fact that the initial rate under zeroth-order conditions was the same, to within experimental error, as the rate of ^{18}O-exchange in a similar solution. It was inferred that the exchange of oxygen occurred *via* heterolysis to the nitronium ion, and that it was the rate of this heterolysis which limited the rates of nitration of reactive aromatic compounds.

$$HNO_3 + H^+ \underset{k_{-1}}{\overset{k_1}{\rightleftharpoons}} NO_2^+ + H_2O$$

$$ArH + NO_2^+ \underset{(\text{For zeroth-order nitration } k_2 \gg k_{-1})}{\overset{k_2}{\rightleftharpoons}} [ArHNO_2^+] \quad \| \quad H_2{}^{18}O + NO_2^+ \rightleftharpoons H^{18}O.NO_2 + H^+$$

In the process of ^{18}O-exchange the nitronium ion mechanism requires that the rate of nitronium ion formation be the rate at which the label

Nitrating systems, A

appears in nitric acid; this must also be equal to the zeroth-order rate of nitration. Some results are given in table 2.2. If, on the other hand, the effective reagent were the nitric acidium ion, there would be no necessary connection between the rates of the two processes, the nitration of an aromatic and the nitration of water. Clearly then, in aqueous nitric acid containing as much as ~ 60 mols. % of water the nitronium ion is still the effective nitrating agent. In the case of 2-mesitylethane-sulphonate anion (v) the zeroth-order rate exceeded the rate of ^{18}O-exchange by about 15 %; the difference was attributed to uncertainties in extrapolating results, and to the possibility of a salt effect on the rate of nitration.

TABLE 2.2   *Zeroth-order rates of nitration and of ^{18}O-exchange in aqueous nitric acid at $0°C$*

Compound	[HNO$_3$] mols %	[ArH] mols %	$10^4 k_0$/mols % s^{-1} Nitration*	Exchange
Mesitylene-α-sulphonate	39·02	0·23	2·12	2·16
anion	38·89	0·32	1·66	1·66
Iso-durene-α^2-sulphonate	39·37	0·27	1·76	1·93
anion	37·39	0·29	1·19	1·26

* Corrected for incompleteness of trapping of the nitronium ion by the sulphonates at the concentrations used.[16]

The rates of nitration of mesitylene-α-sulphonate anion (III) and iso-durene-α^2-sulphonate anion (IV) in mixtures of aqueous nitric and perchloric acid followed a zeroth-order rate law.[15b] Although the rate of exchange of oxygen could not be measured because of the presence of perchloric acid, these results again show that, under conditions most amenable to its existence and involvement, the nitric acidium ion is ineffective in nitration.

Nitrous acid anticatalyses nitration in aqueous nitric acid more strongly than in pure nitric acid (§ 4.3.2).

2.3 NITRATION IN CONCENTRATED SOLUTIONS OF SULPHURIC
ACID

2.3.1 *The state of nitric acid in 98–100% sulphuric acid*

In this section the pioneering work of Hantzsch will several times be
mentioned. That later techniques made it necessary to modify his
conclusions should not be allowed to obscure the great originality of his
approach since investigations using these media provided the most
compelling evidence for the existence of the nitronium ion.

The two absorption bands, at 1050 and 1400 cm^{-1}, which appear in
the Raman spectra of solutions of nitric acid in concentrated sulphuric
acid are not attributable to either of the acid molecules.[17a, 18] In oleum[17b]
the lower band appears at 1075–1095 cm^{-1}. That these bands seemed to
correspond to those in the spectra of anhydrous nitric acid and solid
dinitrogen pentoxide caused some confusion in the assignment of the
spectrum.[19] The situation was resolved by examining the Raman spectra
of solutions of nitric acid in perchloric or selenic acids[18], in which the
strong absorption at 1400 cm^{-1} is not accompanied by absorption at
about 1050 cm^{-1}. Thus, the band at 1400 cm^{-1} arises from the nitronium
ion, and the band at about 1050 cm^{-1} can be attributed in the cases of
nitric acid and solid dinitrogen pentoxide[20, 21] to the nitrate ion formed
according to the following schemes:

$$2HNO_3 \rightleftharpoons NO_2^+ + NO_3^- + H_2O,$$
$$N_2O_5 \rightleftharpoons NO_2^+ + NO_3^-.$$

In sulphuric acid it arises from the bisulphate ion:

$$HNO_3 + 2H_2SO_4 \rightleftharpoons NO_2^+ + H_3O^+ + 2HSO_4^-.$$

The most recent work indicates that in anhydrous sulphuric acid the
above conversion is complete.[18] The slightly modified absorption band
in oleum arises from the hydrogen pyrosulphate ion formed in the
following way:[17b]

$$HSO_4^- + H_2S_2O_7 \rightleftharpoons H_2SO_4 + HS_2O_7^-.$$

Raman spectroscopy provides the easiest way of estimating the concentration of nitronium ions in different media (§2.4.1). The concentration,
determined by infra-red spectroscopy, of nitronium ions in nitric acid
was increased markedly by the addition of sulphuric acid.[10]

The conversion of nitric acid into another species in concentrated
sulphuric acid was shown by the fact that, whereas the ultraviolet

absorption spectrum of nitric acid in 84·5 % sulphuric acid resembled that of absolute nitric acid, a solution in 100% sulphuric acid was virtually transparent.[22a] The ultraviolet spectra of solutions of nitric acid in sulphuric acid will be further discussed later (§2.4.1).

The depression of the freezing point of sulphuric acid by the addition of nitric acid has historically been the subject of confusion. Hantzsch suggested[22] that, because sulphuric acid is the stronger acid, the following equilibria might exist in these solutions:

$$HNO_3 + H_2SO_4 \rightleftharpoons H_2NO_3^+ + HSO_4^-,$$
$$HNO_3 + 2H_2SO_4 \rightleftharpoons H_3NO_3^{2+} + 2HSO_4^-.$$

He observed an i-factor of 3 and argued for the formation of the di-protonated acid. He interpreted the high electrical conductivity of these media[23] in support of this.

The situation has been examined more recently[24] and an i-factor of about 4 has been observed, consistent with the formation of the nitronium ion. The actual value was 3·82, slightly lower than expected because the water formed is not fully protonated.

Nitric acid in oleum ionizes in the following way:[17b, 25]

$$HNO_3 + 2H_2S_2O_7 \rightleftharpoons NO_2^+ + HS_2O_7^- + 2H_2SO_4.$$

Solutions of nitric acid in 100 % sulphuric acid have a high electrical conductivity. If nitric acid is converted into a cation in these solutions, then the migration of nitric acid to the cathode should be observed in electrolysis. This has been demonstrated to occur in oleum and, less conclusively, in concentrated acid,[26] observations consistent with the formation of the nitronium ion, or the mono- or di-protonated forms of nitric acid. Conductimetric measurements confirm the quantitative conversion of nitric acid into nitronium ion in sulphuric acid.[11b]

Related studies have been made using perchloric acid. From mixtures of anhydrous nitric and perchloric acids in the appropriate proportions, Hantzsch[22a] claimed to have isolated two salts whose structures supported his hypothesis concerning the nature of nitric acid in strong mineral acids. He represented the formation of the salts by the following equations:

$$HNO_3 + HClO_4 \longrightarrow (H_2NO_3^+)(ClO_4^-),$$
$$HNO_3 + 2\,HClO_4 \longrightarrow (H_3NO_3^{2+})(ClO_4^-)_2.$$

The salts had a high electrical conductivity, and it was claimed that the values of the molar conductances at infinite dilution showed the formation of a binary and ternary electrolyte respectively.

14

Hantzsch's work has been reinvestigated.[6a] It was found that the readily obtainable product of the composition $(H_3NO_3{}^{2+})(ClO_4{}^-)_2$ could be separated into nitronium perchlorate and hydroxonium perchlorate by fractional recrystallization from nitromethane. The nitronium salt could be obtained pure, and its structure has been determined by X-ray crystallography.[27] Investigation of the Raman spectrum of this compound established unequivocally the existence of the nitronium ion in many of the media used in nitration.[6b] The same workers were not able to prepare a compound of the composition $(H_2NO_3{}^+) (ClO_4{}^-)$. They suggested that because nitric acid is difficult to remove from nitronium perchlorate, Hantszch may have obtained a mixture of the observed composition by chance.

Nitronium salts of many acids have been prepared,[6a, 28] and some are commercially available. They have been used as nitrating agents (§4.4.2).

2.3.2 *Nitration in concentrated solutions of sulphuric acid*

Concentrated solutions are here considered to be those containing $> c.\ 89\%$ by weight of sulphuric acid. In these solutions nitric acid is completely ionised to the nitronium ion. This fact, and the notion that the nitronium ion is the most powerful electrophilic nitrating species,[29] makes operation of this species in these media seem probable. Evidence on this point comes from the effect on the rate of added water (§2.4.2)

A simple kinetic order for the nitration of aromatic compounds was first established by Martinsen for nitration in sulphuric acid[30] (Martinsen also first observed the occurrence of a maximum in the rate of nitration, occurring for nitration in sulphuric acid of 89–90% concentration). The rate of nitration of nitrobenzene was found to obey a second-order rate law, first order in the concentration of the aromatic and of nitric acid. The same law certainly holds (and in many cases was explicitly demonstrated) for the compounds listed in table 2.3.

Although the proportion of nitric acid present as nitronium ions does not change between 90% and 100% sulphuric acid, the rate constants for nitration of most compounds decrease over this range. Fig. 2.1 illustrates the variation with acidity of the second-order rate constants of the nitration of a series of compounds of widely differing reactivities. Table 2.4 lists the results for nitration in 95% and 100% acid of a selection of less completely investigated compounds.

One explanation of this phenomenon was that the nitrating power of a

Nitrating systems, A

solution of nitric acid in sulphuric acid depended on a balance of its acidity, leading to the formation of the nitronium ion, and its basicity which was considered to catalyse the reaction by aiding the loss of the proton.[39] The decrease in the rate above 90 % acid was ascribed to the diminution of the concentration of bisulphate ions, which functioned as the base. This hypothesis was disproved by Melander, who showed that in nitration the loss of the proton was kinetically insignificant (§6.2.2).

TABLE 2.3 *Nitration in concentrated sulphuric acid: compounds which have been studied kinetically*

Compound	Reference
Acetophenone	31
Anilinium ion	36a, 47a
Anthraquinone	32
Substituted anthraquinones	33
Benzenesulphonic acid	30b, 34a
Benzoic acid	30b
2-Chloro-5-nitroaniline	36b
o- and p-Chloronitrobenzene	30b, 34b, 37–8
m-Chloronitrobenzene	37, 30b
N,N-Dimethylaniline N-oxide	31
1,3-Dimethyl-4,6-dinitrobenzene	30b, 35, 37
2,4-Dinitroanisole	30b
Dinitromesitylene	35
2,4-Dinitrophenol	30b
2,4-Dinitrotoluene	39
p-Nitroaniline	36b
Nitrobenzene	30a, 35, 40a
Nitrobenzoic acids	30b
p-Nitrotoluene	38
Nitrogen heteroaromatics	41–44
Compounds with positively charged substituents	45–47

Many aromatic compounds are sufficiently basic to be appreciably protonated in concentrated sulphuric acid. If nitration occurs substantially through the free base, then the reactivity of the conjugate acid will be negligible. Therefore, increasing the acidity of the medium will, by depleting the concentration of the free base, reduce the rate of reaction. This probably accounts for the particularly marked fall in rate which occurs in the nitration of anthraquinone, benzoic acid, benzenesulphonic acid, and some nitroanilines (see table 2.4).

However, this is only a partial explanation, because many of the compounds listed in table 2.4 are not appreciably protonated, even in

16

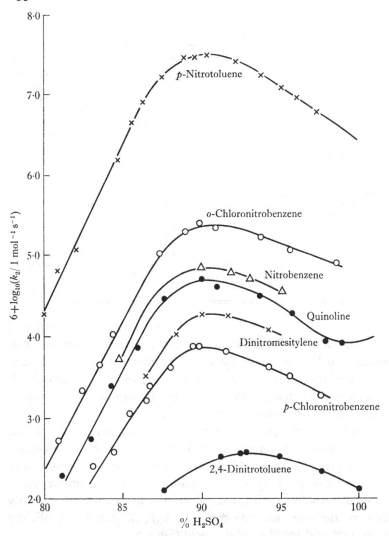

Fig. 2.1. Rate profiles for nitration in 80–100 % sulphuric acid.
For references see table 2.3.

oleum. For these, the variation in the rate is attributed to the change in activity coefficients which is known to occur in strongly acidic solutions. This should be important for, according to transition-state theory the rate should be given by the following expression (in which $f_{\ddagger}$ is the activity coefficient of the transition state):

$$\text{rate} = k \,[\text{NO}_2{}^+]\,[\text{ArH}]\,(f_{\text{NO}_2{}^+} \cdot f_{\text{ArH}}/f_{\ddagger}).$$

TABLE 2.4   *Second-order rate constants for nitration at 25 °C in 95 and 100% sulphuric acid*

Compound	Ref.	$k_2/\text{l mol}^{-1}\,\text{s}^{-1}$		Ratio $\dfrac{k_2\,(95\%)}{k_2\,(100\%)}$	Protonated in 100% sulphuric acid (%)
		95%	100%		
Anilinium	47a	1·5	$6\cdot6\times10^{-2}$	2·2	.
1,3-Dimethyl-4,6-dinitrobenzene	30b	$6\cdot7\times10^{-5}$	$2\cdot3\times10^{-5}$	2·9	.
2,4-Dinitroanisole	30b	$2\cdot8\times10^{-3}$	$8\cdot8\times10^{-4}$	3·2	.
2,4-Dinitrophenol	30b	$1\cdot4\times10^{-2}$	$6\cdot5\times10^{-3}$	2·2	.
o-Chloronitrobenzene	30b	$1\cdot2\times10^{-1}$	$3\cdot6\times10^{-2}$	3·3	.
m-Chloronitrobenzene	30b	$6\cdot5\times10^{-3}$	$2\cdot3\times10^{-3}$	2·8	.
p-Chloronitrobenzene	30b	$3\cdot0\times10^{-3}$	$8\cdot3\times10^{-4}$	3·6	.
p-Nitroaniline	36b	$3\cdot1\times10^{-3}$	$1\cdot8\times10^{-4}$	~17	.
Nitrobenzene	30b	$2\cdot5\times10^{-2}$	$6\cdot2\times10^{-3}$	4·05	28
Anthraquinone	32	$6\cdot2\times10^{-4}$	$8\cdot8\times10^{-5}$	7	99·65
Benzenesulphonic acid	30b	$4\cdot3\times10^{-1}$	$3\cdot9\times10^{-2}$	~12	.
Benzoic acid	30b	>1·7	$9\cdot0\times10^{-2}$	>19	99·98

Investigations of the solubilities of aromatic compounds in concentrated and aqueous sulphuric acids showed the activity coefficients of nitro-compounds to behave unusually when the nitro-compound was dissolved in acid much more dilute than required to effect protonation.[48] This behaviour is thought to arise from changes in the hydrogen-bonding of the nitro group with the solvent.

The activity coefficients in sulphuric acid of a series of aromatic compounds have been determined.[49] The values for three nitro-compounds are given in fig. 2.2. The nitration of these three compounds over a wide range of acidity was also studied,[38] and it was shown that if the rates of nitration were corrected for the decrease of the activity coefficients, the corrected rate constant, k_2/f_{ArH}, varied only slightly between 90% and 100% sulphuric acid (fig. 2.3).

The value of the second-order rate constant for nitration of benzenesulphonic acid in anhydrous sulphuric acid varies with the concentration of the aromatic substrate and with that of additives such as nitromethane and sulphuryl chloride.[34a] The effect seems to depend on the total concentration of non-electrolyte, moderate values of which (up to about 0·5 mol l^{-1}) depress the rate constant. More substantial concentrations of non-electrolytes can cause marked rate enhancements in this medium.[34b] Added hydrogen sulphate salts or bases such as pyridine which generate hydrogen sulphate salts in anhydrous sulphuric acid, at concentrations up to 3 mol l^{-1} accelerate the nitration of 1-chloro-4-nitrobenzene and of the phenyltrimethylammonium ion.[34b] Similar effects have been observed for nitration in concentrated aqueous acid.[35] Unfortunately no information is available about the effect which these additives have on the activity coefficients of aromatic substrates.

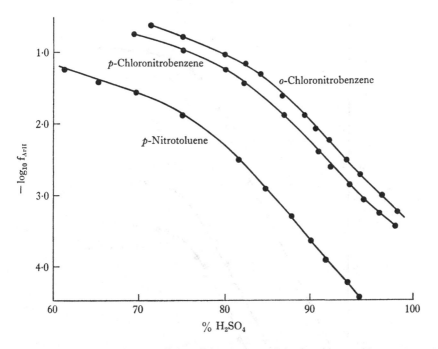

Fig. 2.2. Activity coefficients[49] in aqueous sulphuric acid at 25 °C.

2.4 NITRATION IN AQUEOUS SOLUTIONS OF MINERAL ACIDS

2.4.1 *The state of nitric acid in aqueous sulphuric acid*

Nitric acid is completely converted into nitronium ions in concentrated sulphuric acid (§2.3.1):

$$HNO_3 + 2H_2SO_4 \rightleftharpoons NO_2^+ + NO_3^- + 2HSO_4^-.$$

Raman spectroscopy [17c, d, 50] showed that the addition of up to 10% of water does not affect the concentration of nitronium ions; further dilution reduces the concentration of this species, which is not detectable in solutions containing < 85% sulphuric acid. The introduction of

water into concentrated sulphuric acid generates bisulphate and hydroxonium ions, which tend to repress the formation of nitronium ions, but
the above equilibrium so much favours this species that even moderate
quantities of water have a negligible effect on its concentration.

The ultraviolet spectra of solutions of potassium nitrate in various
concentrations of sulphuric acid have been studied,[51] and absorptions

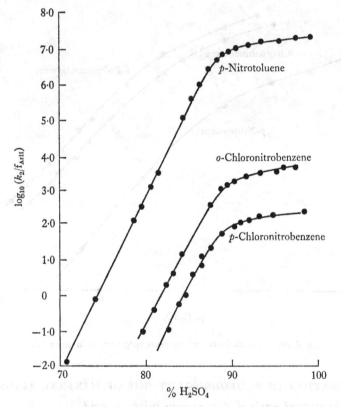

Fig. 2.3. Rate profiles for nitration at 25 °C corrected for variation
in activity coefficients.[38]

arising from the nitrate ion, molecular nitric acid and the nitronium
ion have been observed. The estimation of these species was made
somewhat indirectly because the absorptions of the nitrate ion and
molecular nitric acid were so similar, and because the maximum in the
absorption of the nitronium ion was inaccessible to measurement. The
results show that in concentrations up to 15 % sulphuric acid the nitrate

ion exists exclusively. In the concentration range 15–70% sulphuric acid the nitrate ion and molecular nitric acid coexist, with the concentration of the latter species becoming dominant at higher acidities. Between 72% and 82% sulphuric acid molecular nitric acid is the sole species and in 89% sulphuric acid the ionisation to nitronium ion is complete. The shift in the spectroscopic absorption of the solution between 89–98% sulphuric acid has led to the idea that the species present at the lower acidity is the nitric acidium ion which, in the more concentrated solution, is dehydrated to yield the nitronium ion. This hypothesis is difficult to refute from the data of ultraviolet spectroscopy alone, but the absence of absorptions assignable to the nitric acidium ion in the Raman spectrum, and the presence of bands due to the nitronium ion, with intensities which do not vary over this range of acidity, is evidence against the formation of appreciable concentrations of nitric acidium ion.[51]

The vapour pressure of nitric acid, over solutions in sulphuric acid, reaches a maximum with 84·5% sulphuric acid, the acidity corresponding to the formation of the monohydrate.[52]

2.4.2 *Nitration in aqueous sulphuric acid*

Addition of water to solutions of nitric acid in 90% sulphuric acid reduces rates of nitration. Between 90% and 85% sulphuric acid the decrease in rate parallels the accompanying fall in the concentration of nitronium ions.[35] This is good evidence for the operation of the nitronium ion as the nitrating agent, both in solutions more acidic than 90% and in weakly diluted solutions in which nitronium ion is still spectroscopically detectable.

As the medium is still further diluted, until nitronium ion is not detectable, the second-order rate coefficient decreases by a factor of about 10^4 for each decrease of 10% in the concentration of the sulphuric acid (figs. 2.1, 2.3, 2.4). The active electrophile under these conditions is not molecular nitric acid because the variation in the rate is not similar to the corresponding change in the concentration of this species, determined by ultraviolet spectroscopy or measurements of the vapour pressure.[51-2]

The continued effectiveness of the nitronium ion in relatively dilute solutions has been indicated by comparing the dependence of the rates on the concentration of sulphuric acid, with the acidity-dependence of the ionisation of model compounds. The H_R (formerly J_0 or C_0) acidity scale [53, 54a, c] is determined from the ionisation of tri-aryl carbinols, according to the following scheme:

$$XOH + H^+ \rightleftharpoons X^+ + H_2O.$$

The H_0 acidity function[55] relates to indicators ionising according to the different scheme:

$$B + H^+ \rightleftharpoons BH^+.$$

If it be assumed that the ionising characteristics of nitric acid are similar to those of the organic indicators used to define the scales of acidity, then a correspondence between the acidity-dependence of nitration and H_R would suggest the involvement of the nitronium ion, whereas a correspondence with H_0 would support the hypothesis that the nitric acidium ion were active. The analogies with H_R and H_0 are expressed in the first and last pairs of the following equations respectively. The symbol AQ represents anthraquinone, the indicator originally used in this way for comparison with the acidity dependence of the rate of nitration of nitrobenzene:[35]

$$HNO_3 + 2H_2SO_4 \rightleftharpoons NO_2^+ + H_3O^+ + 2HSO_4^-,$$

$$Ar_3COH + 2H_2SO_4 \rightleftharpoons Ar_3C^+ + H_3O^+ + 2HSO_4^-,$$

$$HNO_3 + H_2SO_4 \rightleftharpoons H_2NO_3^+ + HSO_4^-,$$

$$AQ + H_2SO_4 \rightleftharpoons AQH^+ + HSO_4^-.$$

There is increasing evidence that the ionisation of the organic indicators of the same type, and previously thought to behave similarly, depends to some degree on their specific structures, thereby diminishing the generality of the derived scales of acidity.[56] In the present case, the assumption that nitric acid behaves like organic indicators must be open to doubt. However, the H_R and H_0 scales are so different, and the correspondence of the acidity-dependence of nitration with H_R so much better than with H_0, that the effectiveness of the nitronium ion is firmly established.[35] The relationship between rates of nitration and H_R was subsequently shown to hold up to about 82 % sulphuric acid for nitrobenzene, p-chloronitrobenzene, phenyltrimethylammonium ion, and p-tolyltrimethylammonium ion,[53] and for various other compounds.[54b]

Although the difference between the H_0 and H_R scales is sufficient to permit such a gross mechanistic distinction to be made, the acidity-dependence of nitration deviates from a close correspondence with the

H_R function.[53, 54b] A better correlation, up to nearly 89% sulphuric acid, is obtained by comparing the results at 25 °C with the acidity function $- (H_R + \log a_{H_2O})$.[31, 42a, 43a] In these comparisons a straight line of approximately unit slope is obtained (fig. 2.4), although for the nitration of benzene in acidities greater than 68% sulphuric acid,[57a] the slope becomes 1·20 (fig. 2.5).

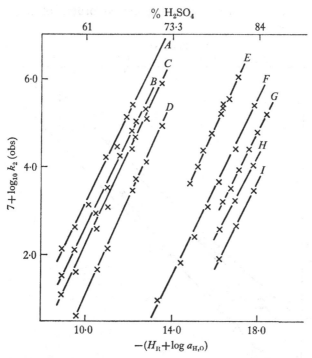

Fig. 2.4. Graph of $[7 + \log_{10} k_2 \text{(obs)}/\text{l mol}^{-1} \text{s}^{-1}]$ against $- (H_R + \log a_{H_2O})$. A, Benzene; B, fluorobenzene; C, bromobenzene; D, chlorobenzene (the ordinate in this case is $6 + \log k_2$); E, tri-methyl-*p*-tolylammonium ion; F, isoquinolinium ion and *N*-methyliso-quinolinium ion; G, benzonitrile; H, quinolinium ion: I, *N*-methylquinolinium ion.

The correlations of rates with acidity functions provide a convenient means of treating results, and their uses will frequently be illustrated. However, their status is empirical, for whilst the acidity dependence of nitration becomes less steep with increasing temperature, the slope of $- H_R$ increases.[53, 57a].

Fig. 2.4, illustrates the variation with the concentration of sulphuric acid of the logarithm of the second-order rate coefficients for the nitration of a series of compounds for which the concentration of effective aromatic does not change with acidity. Such plots, which in the range of acidity 75–85 % sulphuric acid give slopes

$$d\,(\log k_2)/d\,(\%\,\mathrm{H_2SO_4}) \sim 0{\cdot}3{-}0{\cdot}4$$

at 25 °C (§ 8.2.1) are, as will be seen from what has just been said, really plots involving acidity rather than specific acid composition.[35] The nitration of benzene in 51–82 % sulphuric acid at 25 °C has been studied, and rate coefficients spanning eleven orders of magnitude determined.[54b, 57a]

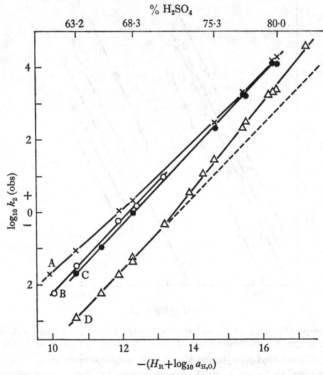

Fig. 2.5. Nitration at 25 °C in aqueous sulphuric acid. $[7 + \log_{10} k_2\ (\text{obs})/1\ \text{mol}^{-1}\ \text{s}^{-1}]$ A, Mesitylene × ; B, naphthalene (○); C, toluene (●); D, benzene (△).

That the rate profiles are close to parallel shows that the variations in rates reflect the changing concentration of nitronium ions, rather than idiosyncrasies in the behaviour of the activity coefficients of the aromatic compounds. The acidity-dependences of the activity coefficients of *p*-nitrotoluene, *o*- and *p*-chloronitrobenzene (fig. 2.2, §2.3.2), are fairly shallow in concentrations up to about 75 %, and seem to be parallel. In more concentrated solutions the coefficients change more rapidly and it is to be expected that differences between compounds will become more marked. Thus, the activity coefficients of p-nitrotoluene and o-chloronitrobenzene differ by factors of about 10, 22 and 42 in 75 %, 82 % and 87 % sulphuric acid respectively.[49]

Unfortunately, insufficient data make it impossible to know whether the activity coefficients of all aromatic compounds vary slightly, or whether certain compounds, or groups of compounds, show unusual behaviour. However, it seems that slight variations in relative rates might arise from these differences, and that comparisons of reactivity are less sound in relatively concentrated solutions.

Another reason for treating with caution the results for benzene in solutions more acidic than 68 % is discussed below (§2.5). The acidity-dependences of rates of nitration at 25 °C have been established for the compounds listed in table 2.5.

If the concentration of effective aromatic species does vary with acidity, as sometimes happens if the compound is substantially protonated, then the acidity-dependence of the rate will be less steep than usual, because the concentration of the active free base diminishes significantly with increasing acidity. This situation has been observed in certain cases (§8.2). The fall in the concentration of the active species can be allowed for from a knowledge of its pK_a and the acidity function which, for the particular compound, gives the best measure of the acidity of the medium. Then the corrected acidity-dependence of the rate resembles that observed with compounds the concentration of which does not change significantly with acidity. The nitration of minor species is discussed later (§8.2).

Compounds more reactive than benzene are considered below (§2.5).

2.4.3 *Nitration in aqueous perchloric acid*

Practical difficulties in using concentrated (> 72 % perchloric acid) and the fact that the H_R function is known only up to 60 % perchloric acid,[54c] reduce the value of these media for the study of nitration.

Rates of nitration in perchloric acid of mesitylene, naphthalene and phenol (57·1–61·1 %), and benzene (57·1–64·4 %) have been determined.[57a] The activated compounds are considered below (§2.5). A plot of the logarithms of the second-order rate coefficients for the nitration of benzene against $-(H_R + \log a_{H_2O})$, over the range of acidity for which values of the latter function are available, [54c, 61] yields a straight line of unit slope.[57a]

TABLE 2.5 *Some compounds for which the dependences of rate of nitration upon acidity, in aqueous sulphuric acid at 25 °C, have been determined*

Compound	Acidity range (%)	Reference
Anilinium ion	82–89·5	36*a*
Benzamidium ion	81·2–87·8	31
Benzene	51–82	54*b*, 57*a*
Benzonitrile	80–87	54*b*
Benzylammonium ion	78–80	47*b*
Benzylidyne trifluoride	75–84	57*b*
Benzyltrimethylammonium ion	78–80	47*b*
Bromobenzene*	57–72	54*b*
Chlorobenzene*	57–72	54*b*
2-Chloro-4-nitroaniline	85–>90	36*b*
o-Chloronitrobenzene	80–>90	38
p-Chloronitrobenzene	80–>90	38
Dinitromesitylene	85–>90	35
Fluorobenzene*	54–70	54*b*
Mesitylene	56–80	57*a*
Naphthalene	61–71	57*a*
p-Nitroaniline	84–>90	36*b*
Nitrobenzene	80–>90	30*a*, 35
m-Nitrotoluene	72–87	59
p-Nitrotoluene	71–>90	38
Phenyldimethylhydroxylammonium ion	84·4–>90	31
2-Phenylethyltrimethylammonium ion	63–77	47*c*
3-Phenylpropyltrimethylammonium ion	61–68	47*c*
Phenyltrimethylammonium ion	82–>90	40*b*
Phenyltrimethylstibonium ion	76–82	47*d*
Toluene	63–80	57*a*
p-Tolytrimethylammonium ion	75–82	60
Heterocyclic compounds		
Substituted Benzimidazolium ions	Various	58*a, b*
1-Hydroxy-2,6-dimethoxypyridinium	80–85	44*c*
2-Hydroxyisoquinolinium ion	76–83	43*a*
1-Hydroxy-2-phenylpyridinium	74–79	64
1-Hydroxyquinolinium ion†	82–85	43*b*
Imidazolium ion	83–>90	41*b*
Isoquinolinium ion	71–84	42*a*
2-Methoxyisoquinolinium ion	76–83	43*b*
2-Methylisoquinolinium ion	76–83	42*a*
1-Methylquinolinium	79–83	42*a*
2-Phenylpyridinium	77–81	64
Quinolinium ion	78–>90	41*a*, 42*a*

* For data on these compounds and on a number of polyhalogenobenzenes see Ref. 63, ch. 2.
† Also the free base, quinoline 1-oxide.

2.5 NITRATION AT THE ENCOUNTER RATE IN AQUEOUS SULPHURIC AND PERCHLORIC ACIDS

Second-order rate coefficients for nitration in sulphuric acid at 25 °C fall by a factor of about 10^4 for every 10 % decrease in the concentration of the sulphuric acid (§2.4.2). Since in sulphuric acid of about 90 % concentration nitric acid is completely ionised to nitronium ions, in 68 % sulphuric acid $[NO_2^+] \sim 10^{-8} [HNO_3]$. The rate equation can be written in two ways, as follows:

$$\text{rate} = k_2 [\text{ArH}][\text{HNO}_3] = k_2' [\text{ArH}][\text{NO}_2^+],$$

and from the observed value[54b] of k_2, the value of the true second-order rate coefficient, k_2', for the nitration of benzene at 25 °C in 68 % sulphuric acid follows as $c. \ 4 \times 10^6 \ \text{l mol}^{-1} \ \text{s}^{-1}$. This value approaches quite closely the value ($c. \ 6 \times 10^8 \ \text{l mol}^{-1} \ \text{s}^{-1}$) calculated from the equation[*] $k_2 = 8RT/3\eta$ for the bimolecular rate coefficient for encounter between two species in sulphuric acid under these conditions.[62]

This consideration prompted an investigation of the nitration of benzene and some more reactive compounds in aqueous sulphuric and perchloric acids, to establish to what extent the reactions of these compounds were affected by the speed of diffusion together of the active species.[57a]

In both media a limit was reached beyond which the introduction of further activating substituents did not increase the rate of nitration; this limit was identified as the rate of encounter of the nitronium ions and the aromatic molecules.

The phenomenon was established firmly by determining the rates of reaction in 68·3 % sulphuric acid and 61·05 % perchloric acid of a series of compounds which, from their behaviour in other reactions, and from predictions made using the additivity principle (§9.2), might be expected to be very reactive in nitration. The second-order rate coefficients for nitration of these compounds, their rates relative to that of benzene and, where possible, an estimate of their expected relative rates are listed in table 2.6.

The rates of nitration, under a variety of conditions (56–80 % sulphuric acid, 57–62 % perchloric acid), of mesitylene and benzene

---

[*] The equation is sometimes written, $k_2 = 8RT/3000 \ \eta$. The form preferred here is more correct because it makes no presupposition about the units used. Arithmetical adjustments will normally be necessary to give rate constants in the required units.

were determined. The relative rate varied, being about 70, 36 and 6 in 63 %, 68·3 % and 80 % sulphuric acid respectively. The decrease in this ratio is an expected result of the increasing viscosities of the media, but the factors which affect the threshold of control by diffusion are not fully understood (§3.3).

TABLE 2.6 *Second-order rate coefficients and relative rates for nitration at 25·0 °C in 68·3 % sulphuric acid and 61·05 % perchloric acid*[57a]

Compound	68·3 % sulphuric acid		61·05 % perchloric acid		Estimated rel. rate
	k_2/l mol^{-1} s^{-1}	Rel. rate	k_2/l mol^{-1} s^{-1}	Rel. rate	
Benzene	$5·8 \times 10^{-2}$	1	$8·3 \times 10^{-2}$	1	1
Toluene	1·0	17	1·6	19	23
Biphenyl	$9·2 \times 10^{-2}$	16	.	.	35
o-Xylene	2·2	38	.	.	60
m-Xylene	2·2	38	.	.	400
p-Xylene	2·2	38	7·0	85	50
Mesitylene	2·1	36	6·5	78	16 000
Naphthalene	1·6	28	2·2	27	300
2-Methylnaph-thalene	1·6	28	4·7	56	.
1-Methoxynaph-thalene	2·0	35	7·3	88	.
Phenol	1·4	24	2·6	31	~10000
m-Cresol	.	.	4·9	59	.
Thiophen	.	.	4·3	52	.
1-Naphthol	.	.	7·0	85	.

The possibility that the rate of reaction of benzene is affected by the phenomenon of reaction at the encounter rate is a matter of importance, because benzene is the datum relative to which comparisons of reactivity are made. Up to 68 % sulphuric acid the slope of a plot of log (k_2/l mol^{-1} s^{-1}) against $-(H_R + \log a_{H_2O})$ is unity for data relating to 25 °C, and the rate of encounter is sufficiently great to make it likely that the rate for benzene is unaffected. In solutions more acidic than 68 % sulphuric acid the slope of the plot becomes 1·20; furthermore, the rate for benzene approaches closely the rate of encounter. Both of these points indicate that in the acidity range 68–82 % sulphuric acid comparisons must be made with care, because the observed relative rates, particularly at the higher acidities, may not be true measures of relative reactivity. The situation is illustrated in fig. 2.5. Nitration of reactive compounds at the encounter rate is implicit in the results for the nitration of some aryl-alkanesulphonic acids already discussed (§2.2.4) and in some of the experiments of Deno and Stein.[54b] The implications of reaction at the encounter rate as it affects the study of aromatic reactivity through nitration are further discussed later (§7.1.2).

The results in table 2.6 show that the rates of reaction of compounds such as phenol and 1-napthol are equal to the encounter rate. This observation is noteworthy because it shows that despite their potentially very high reactivity these compounds do not draw into reaction other electrophiles, and the nitronium ion remains solely effective. These particular instances illustrate an important general principle: *if by increasing the reactivity of the aromatic reactant in a substitution reaction, a plateau in rate constant for the reaction is achieved which can be identified as the rate constant for encounter of the reacting species, and if further structural modifications of the aromatic in the direction of further increasing its potential reactivity ultimately raise the rate constant above this plateau, then the incursion of a new electrophile must be admitted.*

REFERENCES

1. Euler, J. (1903). *Justus Liebigs Annln Chem.* **330**, 286.
2. Gillespie, R. J., Hughes, E. D. & Ingold, C. K. (1950). *J. chem. Soc.* p. 2552.
3. Lewis, T. J. (1954). Ph.D. thesis, University of London.
4. Lee, W. H. & Millen, D. J. (1956). *J. chem. Soc.* p. 4463.
5. (a) Fénéant, S. & Chédin, J. (1947). *C. r. hebd. Séanc. Acad. Sci., Paris* **224**, 1008.
   (b) Angus, W. R. & Leckie, A. H. (1935). *Proc. R. Soc. Lond.* A **149**, 327.
   (c) Redlich, O. & Nielsen, L. E. (1943). *J. Am. chem. Soc.* **65**, 654.
6. (a) Goddard, D. R., Hughes, E. D. & Ingold, C. K. (1950). *J. chem. Soc.* p. 2559.
   (b) Millen, D. J. (1950). *J. chem. Soc.* p. 2606.
7. Susz, B. & Briner, E. (1935). *Helv. chim. Acta* **18**, 378.
8. Chédin, J. (1935). *C. r. hebd. Séanc. Acad. Sci., Paris* **201**, 552.
9. Ingold, C. K. & Millen, D. J. (1950). *J. chem. Soc.* p. 2612.
10. Marcus, R. A. & Frescoe, J. M. (1957). *J. chem. Phys.* **27**, 564.
11. Gillespie, R. J. (a) (1950), *J. chem. Soc.* p. 2493; (b) with Wasif, S. (1953). *J. chem. Soc.* p. 221.
12. Chédin, J. & Fénéant, S. (1947). *C. r. hebd. Séanc. Acad. Sci., Paris* **224**, 930.
13. Hughes, E. D., Ingold, C. K. & Reed, R. I. (1950). *J. chem. Soc.* p. 2400.
14. Halberstadt, E. S., Hughes, E. D. & Ingold, C. K. (1950). *J. chem. Soc.* p. 2441.
15. Bunton, C. A. (a) with Halevi, E. A. & Llewellyn, D. R. (1952), *J. chem. Soc.* p. 4913; (b) with Halevi, E. A. (1952), *J. chem. Soc.* p. 4917; (c) with Stedman, G. (1958), *J. chem. Soc.* p. 2420.

References

16. Ingold, C. K. (1959). *Substitution at Elements other than Carbon.* Jerusalem. The Wiezmann Science Press of Israel.
17. Chédin, J. (a) (1935). *C. r. hebd. Séanc. Acad. Sci., Paris* **200**, 1397.
    (b) (1936). *C. r. hebd. Séanc. Acad. Sci., Paris* **202**, 220.
    (c) (1937). *Annls Chim.* **8**, 243.
    (d) (1944). *Mém. Services chim. État* **31**, 113.
18. Ingold, C. K., Millen, D. J. & Poole, H. G. (1950). *J. chem. Soc.* p. 2576.
19. Gillespie, R. J. & Millen, D. J. (1948). *Q. Rev. chem. Soc.* **2**, 277.
20. Chédin, J. & Pradier, J. C. (1936). *C. r. hebd. Séanc. Acad. Sci., Paris* **203**, 722.
21. Grison, E., Ericks, K. & de Vries, J. L. (1950). *Acta crystallogr.* **3**, 290.
22. Hantzsch, A. (a) (1925). *Ber. dt. chem. Ges.* **58**, 941; (1930) *Z. phys. Chem.* **149**, 161.
    (b) (1908), *Z. phys. Chem.* **61**, 257; *Z. phys. Chem.* **62**, 178, 626; (1909), *Z. phys. Chem.* **65**, 41; (1910), *Z. phys. Chem.* **68**, 204.
23. (a) Saposchnikow, A. (1904), *Z. phys. Chem.* **49**, 697; (1905), *Z. phys. Chem.* **51**, 609; (1905), *Z. phys. Chem.* **53**, 225.
    (b) Bergius, F. (1910). *Z. phys. Chem.* **72**, 338.
24. Gillespie, R. J., Graham, J., Hughes, E. D., Ingold, C. K. & Peeling, E. R. A. (1950). *J. chem. Soc.* p. 2504.
25. Millen, D. J. (1950). *J. chem. Soc.* p. 2589.
26. Bennett, G. M., Brand, J. C. D. & Williams, G. (1946). *J. chem. Soc.* p. 875.
27. Cox, E. G., Jeffrey, G. A. & Truter, M. R. (1948). *Nature, Lond.* **162**, 259.
28. Olah, G. A. & Kuhn, S. J. (1964). *Friedel–Crafts and Related Reactions,* vol. 3 (ed. G. A. Olah), p. 1427. New York: Interscience.
29. Ingold, C. K. (1953). *Structure and Mechanism in Organic Chemistry,* p. 282. London: Bell.
30. Martinsen, H. (a) (1905). *Z. phys. Chem.* **50**, 385.
    (b) (1907). *Z. phys. Chem.* **59**, 605.
31. Moodie, R. B., Penton, J. R. & Schofield, K. (1969). *J. chem. Soc.* B, p. 578.
32. Lauer, L. & Oda, R. (1936). *J. prakt. Chem.* **144**, 176.
33. Oda, R. & Ueda, U. (1941). *Bull. Inst. Phys. Chem. Res. (Tokyo)* **20**, 335.
34. (a) Surfleet, B. & Wyatt, P. A. H. (1965). *J. chem. Soc.* p. 6524. Akand, M. A. & Wyatt, P. A. H. (1967). *J. chem. Soc.* B, p. 1326.
    (b) Bonner, T. G. & Brown, F. (1966). *J. chem. Soc.* B, p. 658.
35. Westheimer, F. H. & Kharasch, M. S. (1946). *J. Am. chem. Soc.* **68**, 1871.
36. Hartshorn, S. R. & Ridd, J. H. (1968). *J. chem. Soc.* B (a) p. 1063; (b) p. 1068.
37. Klemenc, A. & Schöller, R. (1924). *Z. anorg. Chem.* **141**, 231.
38. Vinnik, M. I., Grabovskaya, Zh. E. & Arzamaskova, L. N. (1967). *Russ. J. phys. Chem.* **41**, 580.
39. Bennett, G. M., Brand, J. C. D., James, D. M., Saunders, T. G. & Williams G. (1947). *J. chem. Soc.* p. 474.
40. Bonner, T. G., Bowyer, F. & Williams, G.: (a) (1953). *J. chem. Soc.* p. 2650; (b) (1952). *J. chem. Soc.* p. 3274.
41. (a) Austin, M. W. & Ridd, J. H. (1963). *J. chem. Soc.* p. 4204.
    (b) Austin, M. W., Blackborow, J. R., Ridd, J. H. & Smith, B. V. (1965). *J. chem. Soc.* p. 1051.

42. Moodie, R. B. & Schofield, K. (*a*) with Williamson, M. J. (1964). *Nitro-Compounds*, p. 89. *Proceedings of International Symposium, Warsaw* (1963). London: Pergamon Press. (*b*) with Qureshi, E. A. & (in part) Gleghorn, J. T. (1968). *J. chem. Soc.* B, p. 312.

43. Gleghorn, J. T., Moodie, R. B. & Schofield, K. (*a*) with Williamson, M. J. (1966). *J. chem. Soc.* B, p. 870; (*b*) with Qureshi, E. A. (1968). *J. chem. Soc.* B, p. 316.

44. Johnson, C. D. & Katritzky, A. R. (*a*) with Ridgewell, B. J. & Viney, M. (1967). *J. chem. Soc.* B, p. 1204; (*b*) with Viney, M. (1967). *J. chem. Soc.* B, p. 1211; (*c*) with Shakir, N. & Viney, M. (1967). *J. chem. Soc.* B, p.1213.

45. Utley, J. H. P. & Vaughan, T. A. (1968). *J. chem. Soc.* B, p. 196.

46. Gilow, H. M. & Walker, G. L. (1967). *J. org. Chem.* 32, 2580.

47. (*a*) Brickmann, M. & Ridd, J. H. (1965). *J. chem. Soc.* p. 6845.
    (*b*) Brickmann, M., Utley, J. H. P., & Ridd, J. H. (1965). *J. chem. Soc.* p. 6851.
    (*c*) Modro, T. A. & Ridd, J. H. (1968). *J. chem. Soc.* B, p. 528.
    (*d*) Gastaminza, A., Modro, T. A., Ridd, J. H. & Utley, J. H. P. (1968). *J. chem. Soc.* B, p. 534.

48. (*a*) Hammett, L. P. & Chapman, R. P. (1934). *J. Am. chem. Soc.* 56, 1282.
    (*b*) Brand, J. C. D. (1950). *J. chem. Soc.* p. 997.
    (*c*) Arnett, E. M., Wu, C. Y., Anderson, J. N. & Bushick, R. D. (1962). *J. Am. chem. Soc.* 84, 1674.

49. Grabovskaya, Zh. E. & Vinnik, M. I. (1966). *Russ. J. phys. Chem.* 40, 1221.

50. Fénéant, S. & Chédin, J. (1955). *Mém. Services chim. État* 40, 292.

51. Bayliss, N. S. & Watts, D. W. (1963). *Aust. J. Chem.* 16, 943.

52. Vandoni, R. (1944). *Mém. Services chim. État* 31, 87.

53. Lowen, A. M., Murray, M. A. & Williams, G. (1950). *J. chem. Soc.* p. 3318.

54. Deno, N. C. (*a*) with Jaruzelski, J. J. & Schriesheim, A. (1955). *J. Am. chem. Soc.* 77, 3044; (*b*) with Stein, R. (1956). *J. Am. chem. Soc.* 78, 578; (*c*) with Berkheimer, H. E., Evans, W. L. & Peterson, H. J. (1959). *J. Am. chem. Soc.* 81, 2344.

55. Hammett, L. P. & Deyrup, A. J. (1932). *J. Am. chem. Soc.* 54, 2721.

56. Challis, B. C. (1965). *A. Rep. chem. Soc.* 62, 249.

57. Coombes, R. G., Moodie, R. B. & Schofield, K. (*a*) (1968). *J. chem. Soc.* B, p. 800; (*b*) (1969). *J. chem. Soc.* B, p. 52.

58. Štěrba, V. & Arient, J. (*a*) with Navrátil, F. (1966). *Colln. Czech. chem. Commun.* 31, 113; (*b*) with Šlosar, J. (1966). *Colln. Czech. chem. Commun.* 31, 1093.

59. Tillett, J. G. (1962). *J. chem. Soc.* p. 1542.

60. Williams, G. & Lowen, A. M. (1950). *J. chem. Soc.* p. 3312.

61. Bunnett, J. F. (1961). *J. Am. chem. Soc.* 83, 4956.

62. Caldin, E. F. (1964). *Fast Reactions in Solution*, p. 10. Oxford: Blackwell.

63. Coombes, R. G., Crout, D. H. G., Hoggett, J. G., Moodie, R. B. & Schofield, K. (1970). *J. chem. Soc.* B, p. 347.

64. Katritzky, A. R. & Kingsland, M. (1968). *J. chem. Soc.* B, p. 862.

3 Nitrating systems:
B. Inert organic solvents

3.1 THE STATE OF NITRIC ACID IN INERT ORGANIC SOLVENTS

The absence of ions in mixtures of acetic acid and nitric acid is shown by their poor electrical conductivity[1] and the Raman spectra of solutions in acetic acid, nitromethane, and chloroform show only the absorptions of the solvent and molecular nitric acid; the bands corresponding to the nitronium and nitrate ions cannot be detected.[2,3,4]

Although no chemical reaction occurs, measurements of the freezing point and infra-red spectra show that nitric acid forms 1:1 molecular complexes with acetic acid[3], ether[5] and dioxan.[5] In contrast, the infra-red spectrum of nitric acid in chloroform and carbon tetrachloride[5,6] is very similar to that of nitric acid vapour, showing that in these cases a close association with the solvent does not occur.

3.2 THE KINETICS OF NITRATION

3.2.1 *Zeroth-order nitrations*

Although the nitronium ion cannot be detected by physical methods in these media, kinetic studies using these solutions have provided compelling evidence for the formation and effectiveness of this species in nitration.

Much of the early work[7] was inconclusive; confusion sprang from the production by the reaction of water, which generally reduced the rate, and in some cases by production of nitrous acid which led to autocatalysis in the reactions of activated compounds. The most extensive kinetic studies have used nitromethane,[8-10] acetic acid,[9] sulpholan,[10] and carbon tetrachloride[6,11] as solvents.

It was from studies of nitration with solutions of nitric acid in nitromethane,[8,9] and later in acetic acid,[9] that Ingold and his co-workers first established the fundamental features of these reactions, and also correctly interpreted them.[9] The use in these experiments of a large excess of nitric acid removed the problem caused by the formation of water.

The rates of nitration of benzene, toluene, and ethylbenzene in solutions of nitric acid (c. 3–7 mol l^{-1}) in nitromethane were independent of the concentration of aromatic compound.[8] These three compounds were nitrated at the same rate, and because the concentration of nitric acid was high compared with that of the aromatic (usually $0 \cdot 05$–$0 \cdot 5$ mol l^{-1}), the reaction proceeded at a constant rate until coming to a sudden stop when all of the aromatic had been consumed. This rate, zeroth order in the concentration of aromatic, is observed in the nitration of a sufficient concentration of a sufficiently reactive compound, and its value is characteristic of the nitration solution. Values of the zeroth-order rate constants for the nitration of benzene at 0 °C are given, with other data, in table 3.1.

Under conditions in which benzene and its homologues were nitrated at the zeroth-order rate, the reactions of the halogenobenzenes ([aromatic] = c. $0 \cdot 1$ mol l^{-1}) obeyed no simple kinetic law. The reactions of fluorobenzene and iodobenzene initially followed the same rates as that of benzene but, as the concentration of the aromatic was depleted by the progress of the reaction, the rate deviated to a dependence on the first power of the concentration of aromatic. The same situation was observed with chloro- and bromo-benzene, but these compounds could not maintain a zeroth-order dependence as easily as the other halogenobenzenes, and the first-order character of the reaction was more marked.

The observation of nitration in nitromethane fully dependent on the first power of the concentration of aromatic was made later.[9] The rate of reaction of p-dichlorobenzene ([aromatic] = $0 \cdot 2$ mol l^{-1}; [HNO_3] = $8 \cdot 5$ mol l^{-1}) obeyed such a law. The fact that in a similar solution 1,2,4-trichlorobenzene underwent reaction according to the same kinetic law, but about ten times slower, shows that under first-order conditions the rate of reaction depends on the reactivity of the compound.

Nitration in acetic acid,[9] in sulpholan[10] and in carbon tetrachloride[6, 11] showed kinetic phenomena similar to those shown in nitromethane; this is significant for it denies nitromethane a chemical involvement in the slow step. (Originally the rate of isomerization of nitromethane to its aci-form was believed to be a factor in the reaction.[8])

The nitration in acetic acid of mesitylene, p-xylene, ethylbenzene and toluene ([aromatic] > 4×10^{-2} mol l^{-1}) was zeroth-order in the concentration of the aromatic compound.[9] Values of the zeroth-order rate constants at 20 °C were $4 \cdot 25 \times 10^{-6}$ mol l^{-1} s^{-1} ([HNO_3] = $5 \cdot 0$ mol l^{-1}) and $6 \cdot 0 \times 10^{-5}$ mol l^{-1} s^{-1} ([HNO_3] = $7 \cdot 0$ mol l^{-1}). The nitration of benzene could be brought under the control of a zeroth-order rate

TABLE 3.1 *Zeroth-order rate constants for nitrations in organic solvents*

Solvent	Temp. °C	Quantity									
Nitromethane[8]	0.0*	$[HNO_3]$/mol l^{-1}	3.5	4.0	4.5	5.0	5.5	6.0	6.5	7.0	
		$10^6\,k_0$/mol l^{-1} s^{-1}	5.8	12.0	20.0	57.5	57.5	135	179	286	
	25.0[13]	$[HNO_3]$/mol l^{-1}	2.0	2.5							
		$10^6\,k_0$/mol l^{-1} s^{-1}	6.8	17.1							
99.8 % acetic acid[9]	20.0†	$[HNO_3]$/mol l^{-1}	5.0	7.0							
		$10^6\,k_0$/mol l^{-1} s^{-1}	4.25	59.6							
Sulpholan[10]	25.0‡	$[HNO_3]$/mol l^{-1}	0.47	0.95	1.91	4.30	4.80	4.91	7.69	7.77	
		$10^6\,k_0$/mol l^{-1} s^{-1}	0.010	0.056	0.43	2.3	3.2	6.3	126	144	
Carbon tetrachloride[6]	25.0	$[HNO_3]$/mol l^{-1}	0.065	0.100	0.150	0.192	0.206	0.229	0.248	0.282	0.301
		$10^6\,k_0$/mol l^{-1} s^{-1}	0.03	0.01	0.04	0.12	0.15	0.28	0.45	1.3	1.5
	39.3	$[HNO_3]$/mol l^{-1}	0.101	0.108	0.192	0.198	0.206	0.292	0.301		
		$10^6\,k_0$/mol l^{-1} s^{-1}	0.03	0.065	0.10	0.11	0.10	0.58	1.1		

* The results are corrected to zero concentration of nitrous acid.
† The kinetic effect of nitrous acid ($[HNO_2] \not> 0.0192$ mol l^{-1}) in the experiment quoted was negligible. The rate constants quoted are average values.
‡ $[HNO_2] \not> 5 \times 10^{-4}$ mol l^{-1}.

law by using high concentrations of aromatic and nitric acid

$$([\text{benzene}] = 3 \cdot 5 \times 10^{-1} \text{ mol } l^{-1} \text{ } [HNO_3] = 12 \cdot 4 \text{ mol } l^{-1}).$$

More typically its reactions showed an intermediacy of kinetic order like that observed with fluorobenzene or iodobenzene in nitromethane.

In acetic acid the rates of nitration of chlorobenzene and bromo-benzene were fairly close to being first order in the concentration of aromatic, and nitration fully according to a first-order law was observed with *o*-, *m*-, and *p*-dichlorobenzene, ethyl benzoate and 1,2,4-trichloro-benzene.

For nitration carried out in sulpholan ($[HNO_3] = 4 \cdot 91 \text{ mol } l^{-1}$), zeroth-order nitration was observed with mesitylene. With toluene and benzene the kinetics were of mixed-order and first-order, respectively.[10]

For nitration in carbon tetrachloride ($[HNO_3] = 0 \cdot 1 - 0 \cdot 2 \text{ mol } l^{-1}$) it has been reported[11] that the reactions of benzene, toluene, *p*-xylene, and mesitylene proceeded according to a zeroth-order kinetic law. The zeroth-order rate constants were claimed to increase according to the sixth power of the concentration of nitric acid, and to decrease by a factor of 300 as the temperature was changed from 0–40 °C, a tempera-ture-dependence corresponding to an Arrhenius exponential factor of − 105 kJ mol^{-1}. Later results[6] differed from these, and it is probable that in the earlier work heterogeneity arose in the nitrating solutions during the course of the reactions because of the formation of water. In these later experiments the nitrations of mesitylene, *o*- and *m*-xylene and toluene ($[\text{aromatic}] > 10^{-3} \text{ mol } l^{-1}$) were of zeroth order in the con-centration of aromatic. The nitration of benzene ($[\text{aromatic}] = 4\text{–}8 \times 10^{-3} \text{ mol } l^{-1}$) was first order, and by using sufficiently low concentrations of aromatic, toluene could also be brought under the control of a first-order law. The values of the zeroth-order rate constants for nitration at 25 °C increased according to the fifth power of the concentration of nitric acid. The rates of reaction were found to decrease threefold in the range of 0–40 °C. The Arrhenius exponential factor was *c.* − 21 kJ mol^{-1}, and did not change as the acidity of the medium was varied from 0·1 to 0·3 mol l^{-1} of nitric acid.

Data for zeroth-order nitration in these various solvents are given in table 3.1. Fig. 3.1 shows how zeroth-order rate constants depend on the concentration of nitric acid, and table 3.2 shows how the kinetic forms of nitration in organic solvents depend on the reactivities of the com-pounds being nitrated.

TABLE 3·2 *Comparison of organic solvents as media for nitration*

	Solvent*		
	Nitromethane[8,9]	99·8% acetic acid[9]	Sulpholan[10]
[HNO$_3$]/mol l^{-1}	5·0	5·0	4·9I
Temperature/°C	0·0	20·0	25·0
k_0/10^{-7} mol l^{-1} s^{-1}	575	42·5	63
Compound nitrated			
Zeroth order	Benzene and homologues	Toluene (benzene)	Mesitylene
Order 0–1	Halogenobenzenes	(Benzene) halogenobenzenes	Toluene
First order	Di- and tri-chlorobenzenes (†)	Di- and tri-chlorobenzenes	Benzene

* The data for nitration in carbon tetrachloride were obtained[6] with much lower concentrations of nitric acid than those tabulated (see table 3.1).
† The data for di-and tri-chlorobenzenes relate to 10·0 and 20·0 °C.

The observation already discussed (§2.2, 2.3, 2.4) of nitrations, in concentrated and aqueous mineral acids and in pure nitric acid, which depend on the first power of the concentration of the aromatic compound, does not help much in elucidating the mechanisms of nitrations under these conditions. In contrast, the observation of zeroth-order

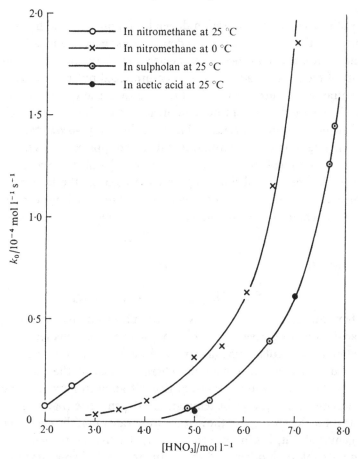

Fig. 3.1. Zeroth-order rates of nitration.

nitration in organic solvents immediately demonstrates that the effective nitrating agent is formed from nitric acid in a slow step. Since proton transfers are unlikely to be slow, the nitrating agent is most probably formed by heterolysis, and must be the nitronium ion. Detailed support for this conclusion comes from the effects of added species on the rate (§§3.2.3, 3.2.4).

Accepting, for the moment without further evidence, that the nitronium ion formed by heterolysis of nitric acid is the active reagent in the solutions under discussion, it remains to consider briefly why nitration in such solutions depends on the concentrations of nitric acid to such high powers (fig. 3.1), and why different solvents behave so differently (table 3.2).

The zeroth-order rates of nitration depend on a process, the heterolysis of nitric acid, which, whatever its details, must generate ions from neutral molecules. Such a process will be accelerated by an increase in the polarity of the medium such as would be produced by an increase in the concentration of nitric acid.[9] In the case of nitration in carbon tetrachloride, where the concentration of nitric acid used was very much smaller than in the other solvents (table 3.1), the zeroth-order rate of nitration depended on the concentration of nitric acid approximately to the fifth power. It is argued therefore that five molecules of nitric acid are associated with a pre-equilibrium step or are present in the transition state. Since nitric acid is evidently not much associated in carbon tetrachloride a scheme for nitronium ion formation might be as follows:

$$5HNO_3 \rightleftharpoons \begin{matrix} NO_2^+ \\ \vdots \\ NO_3^-.HNO_3 \end{matrix} \quad + \quad \begin{matrix} H_3O^+ \\ \vdots \\ NO_3^-.HNO_3 \end{matrix}$$

Here we have the formation of the activated complex from five molecules of nitric acid, previously free, with a high negative entropy change. The concentration of molecular aggregates needed might increase with a fall in temperature in agreement with the characteristics of the reaction already described.[6] It should be noticed that nitration in nitromethane shows the more common type of temperature-dependence (fig. 3.1).

As regards the dependence of the zeroth-order rate upon the character of the organic solvent, it can be seen from fig. 3.1 that acetic acid and sulpholan resemble each other, and that both are 'slower' solvents than nitromethane. The fact is also clear from the magnitudes of the zeroth-order rate constants in these solvents containing the same concentration of nitric acid (table 3.2). As mentioned above, the zeroth-order rate constant would be expected to be greater the more polar the solvent,[9] but it is difficult to give more than qualitative support to this idea because it is unlikely that the macroscopic physical properties of these organic solvents can give a sure guide to polarity. Further, the use of such properties is rendered unsatisfactory by the fact that the substantial concentrations of nitric acid used in these experiments must modify the bulk properties of the organic solvents.

The more strongly acidic a solution of nitric acid at a given concentration is in a particular organic solvent, the more readily should that solvent support zeroth-order nitration. The values of H_0 for solutions of sulphuric acid in nitromethane, sulpholan, and acetic acid show clearly the superiority of nitromethane in this respect.[10]

The relative abilities of nitromethane, sulpholan, and acetic acid to support the ionisation of nitric acid to nitronium ions are closely similar to their efficiencies as solvents in nitration. Raman spectroscopy showed that for a given concentration of mixed acid (1:1 nitric and sulphuric acids) the concentration of nitronium ions in these three solvents varied in the order nitromethane $\gg$ sulpholan $>$ acetic acid. The concentration of mixed acid needed to permit the spectroscopic detection of nitronium ions was 25%, 50% and 60% in the three solvents, respectively[12] (see §4.4.3).

3.2.2 *First-order nitration*

The position reached thus far is summarized by the following relationship. The zeroth-order

$$HNO_3 \underset{b}{\overset{a}{\rightleftharpoons}} NO_2^+ \overset{c}{\longrightarrow} ArNO_2$$

rate depends on process a, and the kinetic order of nitration on what happens to the nitronium ion. If the latter is mainly removed by process c, zeroth-order kinetics will result. If, however, process b accounts mainly for its fate, first-order kinetics will be observed. At a given concentration of nitric acid, process b would increase in importance if either the reactivity of the aromatic compound were decreased, or if in the case of a particular compound its concentration were lowered. The first of these changes is illustrated by the data in table 3.2. The lower reactivity required of the aromatic compound to maintain zeroth-order nitration again brings out the pre-eminence of nitromethane. The deviation of rate away from a zeroth-order towards a first-order dependence upon the concentration of the aromatic is seen in the behaviour of the compounds (table 3.2) which follow an order between zero and unity. This is a consequence of the decrease in the concentration of the aromatic towards the end of the reaction.

With a given aromatic compound a change in the form of the rate law

39

might result if the importance of process *b* were increased by the use of a lower concentration of nitric acid; this would happen because with a lower concentration of nitric acid the polarity of the medium would be decreased, and consequently the rate of process *b* in which ions are converted into neutral molecules would be increased. The nitration of benzene in acetic acid with a high concentration of nitric acid was very nearly of zeroth-order in the concentration of benzene, but with a lower concentration of nitric acid, very nearly of first order.[9] However, this point cannot be considered to be well-established experimentally, because in the experiments quoted the concentration of benzene was altered as well as that of nitric acid, and this could also have led to the observed change in order.

3.2.3 *The effects of added species*

The influence of added species upon the rates and kinetic forms of nitration in organic solvents were of the greatest importance in elucidating details of the processes involved, particularly of the steps leading to the nitronium ion. These influences will first be described, and then in the following section explained. The species to be considered are sulphuric acid, nitrate ions, urea and water. The effect of nitrous acid is considered later (§4.3).

Nitration in organic solvents is strongly catalysed by small concentrations of strong acids; typically a concentration of 10^{-3} mol l^{-1} of sulphuric acid doubles the rate of reaction. Reaction under zeroth-order conditions is accelerated without disturbing the kinetic form, even under the influence of very strong catalysis. The effect of sulphuric acid on the nitration of benzene in nitromethane is tabulated in table 3.3.[9] The catalysis is linear in the concentration of sulphuric acid.

TABLE 3.3 *The effect of sulphuric acid on the zeroth-order nitration of benzene ([benzene] = 0·12 mol l^{-1}) in a solution at $-10\ °C$ of nitric acid ([HNO₃] = 3·0 mol l^{-1}) in nitromethane*

$[H_2SO_4]$/mol l^{-1}	0	5.3×10^{-3}	2.06×10^{-2}	3.29×10^{-2}	4.24×10^{-2}
$10^6 k_0$/mol l^{-1} s^{-1}	1·03	10·5	31·0	47·1	63·8

First-order nitrations in nitromethane are also markedly accelerated by addition of sulphuric acid, the effect being very similar to that observed with zeroth-order reactions.

Unlike the effect of sulphuric acid upon nitration in nitric acid (§2.2.3; where zeroth-order reactions are unknown), the form of the catalysis of zeroth-order nitration in nitromethane by added sulphuric acid does not deviate from a first-order dependence with low concentrations of catalyst.[9]

Salts of strong acids weakly accelerate nitration in organic media; the effect is just detectable in a solution containing 10^{-2} mol l^{-1} of potassium perchlorate. In contrast, nitrate ions strongly and specifically anticatalyse nitration, and they do so without modifying the kinetic order of the reaction. The effect of potassium nitrate on the nitrations of toluene according to a zeroth-order law, and p-dichlorobenzene according to a first-order law in solutions of nitric acid in nitromethane are illustrated in table 3.4. In both cases the reciprocal of the rate is linearly related to the concentration of added nitrate, although with high concentrations (7×10^{-2} mol l^{-1}) of this species the response of the rate deviates from this form. The effect does not depend on the cation associated with the nitrate ion.[9]

TABLE 3.4   *The effects of potassium nitrate on rates of nitration in nitromethane*

Zeroth-order nitration, -10 °C [Toluene] $= 0.09$ mol l^{-1} [HNO$_3$] $= 7.0$ mol l^{-1}		First-order nitration, 20 °C [p-dichlorobenzene] $= 0.1$ mol l^{-1} [HNO$_3$] $= 8.5$ mol l^{-1}	
$\dfrac{[KNO_3]}{\text{mol } l^{-1}}$	$\dfrac{10^6 k_0}{\text{mol } l^{-1}\,s^{-1}}$	$\dfrac{[KNO_3]}{\text{mol } l^{-1}}$	$\dfrac{10^5 k_1}{s^{-1}}$
0	89.3	0	649
3.46×10^{-3}	47.8	10^{-2}	348
5.87×10^{-3}	38.9	2.5×10^{-2}	193
2.32×10^{-2}	21.8	3.5×10^{-2}	164
4.3×10^{-2}	18.7	6.1×10^{-2}	111
1.1×10^{-1}	15.7	8.1×10^{-2}	97
2.8×10^{-1}	13.0	.	.
6.3×10^{-1}	9.4	.	.

In experiments on the nitration of benzene in acetic acid, to which urea was added to remove nitrous acid (which anticatalyses nitration; §4.3.1), the rate was found to be further depressed. The effect was ascribed to nitrate ions arising from the formation of urea nitrate.[9] In the same way, urea depressed the rate of the zeroth-order nitration of mesitylene in sulpholan.[10]

The addition of water depresses zeroth-order rates of nitration, although the effect is very weak compared with that of nitrate ions: concentrations of 6×10^{-1} mol l^{-1} of water, and 4×10^{-3} mol l^{-1} of potassium nitrate halve the rates of reaction under similar conditions. In moderate concentrations water anticatalyses nitration under zeroth-order conditions without changing the kinetic form. This effect is shown below (table 3.5) for the nitration of toluene in nitromethane.[9] More strikingly, the addition of larger proportions of water modifies the kinetic

TABLE 3.5. *The effect of added water on the zeroth-order rates of nitration of toluene ([aromatic] = 0·09 mol l⁻¹) in a solution at −10 °C of nitric acid ([HNO₃] = 7·0 mol l⁻¹) in nitromethane*

$[H_2O]$/mol l^{-1}	0	1.8×10^{-2}	8.1×10^{-2}	1.6×10^{-1}	3.1×10^{-1}	8.6×10^{-1}	1·59
$10^5 k_0$/mol l^{-1} s^{-1}	8·58	8·61	7·76	7·72	5·56	4·11	2·61

form of nitration from a zeroth order to a first-order dependence upon the concentration of the aromatic.* Thus, toluene and *tert*-butylbenzene are nitrated in acetic acid with zeroth-order kinetics, but the addition of 5 % of water drives the kinetics over to the first-order form.[9] Nitrations of a series of compounds in sulpholan were similarly modified by addition of 7·5 % of water, whilst reactions in nitromethane needed about 15 % of water for the transformation to be achieved. The amount of water necessary to effect the change in kinetics depends on the reactivity of the aromatic compound being nitrated; whilst with sulpholan 7·5 % of water was required to remove all of the compounds studied into the regime of the first-order law, toluene, a moderately reactive compound was so removed by $\sim$ 5 % of water.[10] Fig. 3.2 illustrates the effect.

3.2.4 *Discussion of the effects of added species*

The most crucial observation concerning the effects of added species is that nitrate ion anticatalyses nitration without changing the kinetic form of the reaction. This shows that nitrate does not exert its effect by consuming a proportion of the nitronium ion, for, as outlined above, this would tend to bring about a kinetically first-order reaction. Nitrate ions must be affecting the concentration of a precursor of the nitronium

---

* The phenomenon has also been observed in the N-nitration of N-methyl-2,4,6-trinitroaniline and the O-nitration of alcohols.[14]

ion, the precursor being formed in equilibrium with nitric acid. The dependence of the anticatalysis on the first power of the concentration of nitrate indicates that the pre-equilibrium stage is that shown below:

$$2HNO_3 \rightleftharpoons H_2NO_3^+ + NO_3^-.$$

The linear variation of the anticatalysis with the concentration of nitrate, even in very small concentrations, shows that the extent of the autoprotolysis is small.

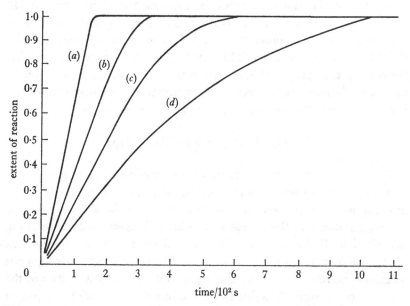

Fig. 3.2. The effect of added water on the rate of nitration of mesitylene. Temperature 25·0 °C. [HNO₃] = 7·73 mol l⁻¹. [Mesitylene] = 0·02 mol l⁻¹. (*a*) [H₂O] = mol l⁻¹. (*b*) [H₂O] = 0·2 mol l⁻¹. (*c*) [H₂O] = 0·5 mol l⁻¹. (*d*) [H₂O] = 1·0 mol l⁻¹.

The nitric acidium ion undergoes slow heterolysis to yield water and the nitronium ion:

$$H_2NO_3^+ \underset{}{\overset{slow}{\rightleftharpoons}} H_2O + NO_2^+.$$

Strong acids catalyse nitration by protonating nitric acid, as shown below:

$$HNO_3 + H_2SO_4 \rightleftharpoons H_2NO_3^+ + HSO_4^-,$$

$$H_2NO_3^+ \underset{}{\overset{slow}{\rightleftharpoons}} H_2O + NO_2^+.$$

In the presence of sulphuric acid this route to the nitric acidium ion is almost entirely dominant, the additional route of protonation *via* the

43

solvent being unlikely in the cases studied. This is shown by the power of the catalysis, so strong that the uncatalysed rate is often negligible in comparison, and by the proportionality of the acceleration to the concentration of sulphuric acid. If the catalysed and uncatalysed pathways to the nitric acidium ion coexisted, the addition of small quantities of sulphuric acid would produce a smaller increase in rate than is found. This phenomenon has been observed in nitration in concentrated nitric acid (§2.2.3).

Neither of the above schemes for forming the nitric acidium ion involves water. However, the addition of moderate quantities of water depresses the zeroth-order rate by up to a factor of four, without disturbing the kinetic form. This last fact shows that an inappreciable fraction of the nitronium ions is reacting with water, and therefore to explain the results it is necessary to postulate the existence of a means, involving water, for the consumption of nitric acidium ions:

$$HNO_3 + H_3O^+ \rightleftharpoons H_2NO_3^+ + H_2O.$$

When large concentrations of water are added to the solutions, nitration according to a zeroth-order law is no longer observed. Under these circumstances, water competes successfully with the aromatic for the nitronium ions, and the necessary condition for zeroth-order reaction, namely that all the nitronium ions should react with the aromatic as quickly as they are formed, no longer holds. In these strongly aqueous solutions the rates depend on the concentrations and reactivities of the aromatic compound. This situation is reminiscent of nitration in aqueous nitric acid in which partial zeroth-order kinetics could be observed only in the reactions of some extremely reactive compounds, capable of being introduced into the solution in high concentrations (§2.2.4).

The outcome of the studies discussed, of nitrations in inert organic solvents, is then the mechanistic scheme formulated below:

$$2HNO_3 \underset{k_{-1}}{\overset{k_1}{\rightleftharpoons}} H_2NO_3^+ + NO_3^-, \tag{1}$$

$$H_2NO_3^+ \underset{k_{-4}}{\overset{k_4}{\rightleftharpoons}} NO_2^+ + H_2O, \tag{4}$$

$$NO_2^+ + ArH \overset{k_5}{\longrightarrow} products. \tag{5}$$

In the presence of catalytic concentrations of sulphuric acid step (1) is superseded by the following equilibrium:

$$H_2SO_4 + HNO_3 \underset{k_{-2}}{\overset{k_2}{\rightleftharpoons}} H_2NO_3^+ + HSO_4^-. \tag{2}$$

And in the presence of substantial concentrations (> 0.2 mol l^{-1}) of water the following scheme assumes some importance:

$$H_3O^+ + HNO_3 \underset{k_{-3}}{\overset{k_3}{\rightleftharpoons}} H_2NO_3^+ + H_2O. \tag{3}$$

Steps (1), (2) and (3), and under certain circumstances (4), are equilibria and have characteristic constants $K_1 \ldots K_4$ respectively.

In the absence of sulphuric acid or much water, zeroth-order conditions will arise when $k_5[\text{ArH}] \gg k_{-4}[\text{H}_2\text{O}]$, in which case the observed rate is given by the following expression:

$$\text{rate} = k_4[\text{H}_2\text{NO}_3^+] = k_4 K_1[\text{HNO}_3]^2 [\text{NO}_3^-]^{-1}.$$

First-order conditions will be observed when $k_{-4}[\text{H}_2\text{O}] \gg k_5[\text{ArH}]$, and then the observed rate is given by the following expression:

$$\text{rate} = k_5[\text{ArH}][\text{NO}_2^+] = k_5 K_1 K_4[\text{ArH}][\text{HNO}_3]^2[\text{NO}_3^-]^{-1}[\text{H}_2\text{O}]^{-1}.$$

Added nitrate anticatalyses nitration under both conditions without affecting the order of the reaction, which is determined solely by the relative magnitudes of $k_5[\text{ArH}]$ and $k_{-4}[\text{H}_2\text{O}]$.

The depletion of the concentration of nitric acidium ion by appreciable quantities of water is expressed by the equilibrium

$$[\text{H}_2\text{NO}_3^+] = K_3[\text{H}_2\text{O}^+][\text{HNO}_3]/[\text{H}_2\text{O}],$$

and provided that the condition $k_5[\text{ArH}] \gg k_{-4}[\text{H}_2\text{O}]$ is maintained the consequence is a reduction in rate without a change in kinetic form. With the addition of so much water that its bulk reactivity towards the nitronium ion exceeds that of the aromatic, the rate will be given by the usual equation, as follows:

$$\text{rate} = k_5[\text{ArH}][\text{NO}_2^+].$$

Like added nitrate, sulphuric acid is not involved in the condition which determines the order of the reaction, and therefore its only effect will be to increase the observed rate constants. The rate under zeroth-order conditions is given by the first of the two expressions below,

45

whereas the rate under first-order conditions is given by the second:

$$\text{rate} = k_4 \, K_2 \, [\text{HNO}_3][\text{H}_2\text{SO}_4][\text{HSO}_4{}^-]^{-1},$$

$$\text{rate} = k_5 \, K_2 \, K_4 \, [\text{ArH}][\text{HNO}_3][\text{H}_2\text{SO}_4][\text{HSO}_4{}^-]^{-1}[\text{H}_2\text{O}]^{-1}.$$

It should be noted that none of the foregoing equations relates to stoichiometric concentrations of additives. Quantitative treatment is precluded by ignorance of the effects of ionic atmosphere and of ion-pairing in these media.

3.3 NITRATION AT THE ENCOUNTER RATE IN INERT ORGANIC SOLVENTS

For nitrations in sulphuric and perchloric acids an increase in the reactivity of the aromatic compound being nitrated beyond the level of about 38 times the reactivity of benzene cannot be detected. At this level, and with compounds which might be expected to surpass it, a roughly constant value of the second-order rate constant is found (table 2.6) because aromatic molecules and nitronium ions are reacting upon encounter. The encounter rate is measurable, and recognisable, because the concentration of the effective electrophile is so small.

A similar circumstance is detectable for nitrations in organic solvents, and has been established for sulpholan, nitromethane, 7·5 % aqueous sulpholan, and 15 % aqueous nitromethane. Nitrations in the two organic solvents are, in some instances, zeroth order in the concentration of the aromatic compound (table 3.2). In these circumstances comparisons with benzene can only be made by the competitive method.* In the aqueous organic solvents the reactions are first order in the concentration of the aromatic (§3.2.3) and comparisons could be made either competitively or by directly measuring the second-order rate constants. Data are given in table 3.6, and compared there with data for nitration in perchloric and sulphuric acids (see table 2.6). Nitration at the encounter rate has been demonstrated in carbon tetrachloride, but less fully explored.[6]

---

* It may seem, at first sight, paradoxical that a competition reaction carried out under conditions in which the measured rate is independent of the concentration of the aromatic can tell us about the relative reactivities of two aromatics. Obviously, the measured rate has nothing to do with the rate of the product-determining step, and what is important in determining relative reactivities is the ratio of the values of k_5 (§ 3.2.4) for two compounds. The criteria to be met for a correct application of the competitive method are well understood.[15]

TABLE 3.6. *Relative rates of nitration at 25·0 °C[10]*

Compound nitrated	From kinetic (K) or competition (C) experiments in:							Estimated relative rates		
	7·5% aqueous sulpholan (K)	Sulpholan (C)	15% aqueous nitromethane (K)	Nitromethane (C)	61·05% perchloric acid (K)*	68·3% sulphuric acid (K)*	68·3% sulphuric acid (C)	†	‡	§
Fluorobenzene	0·091	.	.	0·134||	.	0·105	.	.	.	.
Benzene	1	1	1	1	1	1	1	1	1	1
Biphenyl	15·5	.	.	.	.	16	.	.	.	35
Toluene	20	.	25	21^{16b}	19	17	.	.	.	23
Bromo-mesitylene	30	.	.	.	.	.	.	35	.	.
Naphthalene	33	.	.	.	27	28	.	.	.	300
p-Bromo-anisole	43	.	.	.	.	.	.	110	.	.
o-Xylene	61	.	139	.	.	38	.	40	75	60
m-Xylene	100	.	146	.	.	38	.	540	1200	400
p-Xylene	114	.	130	.	84	38	.	38	68	50
Thiophen	~150	.	.	.	52	.	.	.	.	.
Anisole	175	.	.	.	.	.	.	10000	.	.
2-Methyl-naphthalene	230	.	.	.	57	27·5	.	.	.	.
Mesitylene	350	400±50	400	.	78	36	43±4	176000	56000	16000
1-Methyl-napthalene	450–500	.	.	.	.	.	.	.	.	.
1,6-Dimethyl-napthalene	700	.	.	.	.	.	.	.	.	.
Phenol	700	.	.	.	31	24	.	60000	.	10000

* See table 2.6.
† Calculated from the toluene:benzene ratio for nitration in 7·5% aqueous sulpholan, combined with reported isomer ratios.[12]
‡ Calculated from the toluene:benzene ratio for nitration in 15% aqueous nitromethane and reported isomer ratios.[17]
§ See ref. 17.
\|\| Nitration in acetonitrile.[16]

Nitrating systems, B

It will be seen from table 3.6 that whilst in 68·3 % sulphuric acid the limiting rate of nitration is about 38 times and in 61·05 % perchloric acid about 80 times, in the organic solvents it is 300–400 times the rate of nitration of benzene. Furthermore the limiting rate is very sharply defined for the mineral acids but much more diffuse for the organic solvents. In the mineral acids it seems that complete or partial control by the encounter rate must be operating even for a compound so weakly activated as toluene, whilst with the aqueous organic solvents the limit is reached at the level of reactivity represented roughly by m-xylene. The factors which determine the point of onset of control by encounter have been discussed but are not really understood. One interesting conclusion drawn from the results was that whilst in 68·3 % sulphuric acid the value of the ratio $[NO_2^+]/[HNO_3]$ is about 3×10^{-9}, in the aqueous organic solvents it is about 10^{-13}. The absolute values are doubtful, but these figures do bring out at what very small concentrations the nitronium ion remains the effective nitrating agent.[10]

REFERENCES

1. Kolthoff, I. M. & Willman, A. (1934). *J. Am. chem. Soc.* **56**, 1007.
2. Chédin, J. (1935). *C. r. hebd. Séanc. Acad. Sci., Paris* **201**, 552.
   Chédin, J., Desmaroux, J. & Dalmon, R. (1939). *C. r. hebd. Séanc. Acad. Sci., Paris* **209**, 455.
   Medard, L. & Volkringer, M. (1933). *C. r. hebd. Séanc. Acad. Sci., Paris* **197**, 833.
3. Susz, E., Briner, E. & Favarger, P. (1935). *Helv. chim. Acta* **18**, 375.
4. Taylor, E. G. & Follows, A. G. (1951). *Can. J. Chem.* **29**, 461.
5. Dalmon, R. (1941). *C. r. hebd. Séanc. Acad. Sci., Paris* **213**, 782; (1944). *Mém. Services chim. État.* **31**, 55.
6. Coombes, R. G. (1969). *J. Chem. Soc.* B, p. 1256.
7. Gillespie, R. J. & Millen, D. J. (1948). *Q. Rev. chem. Soc.* **2**, 277.
8. Benford, G. & Ingold, C. K. (1938). *J. chem. Soc.* p. 929.
9. Hughes, E. D., Ingold, C. K. & Reed, R. I. (1950). *J. chem. Soc.* p. 2400.
10. Hoggett, J. G., Moodie, R. B. & Schofield, K. (1969). *J. chem. Soc.* B, p. 1.
11. Bonner, T. G., Hancock, R. A. & Rolle, F. R. (1968). *Tetrahedron Lett.* p. 1665.
    Bonner, T. G., Hancock, R. A., Yousif, G. & (in part) Rolle, F. R. (1969). *J. chem. Soc.* B, p. 1237.
12. Olah, G. A., Kuhn, S. J., Flood, S. H. & Evans, J. C. (1962). *J. Am. chem. Soc.* **84**, 3687.
13. Hughes, E. D., (1959). *Theoretical Organic Chemistry.* International Union of Pure and Applied Chemistry: Section of Organic Chemistry. (The Kekulé Symposium.) London: Butterworths.

14. Hughes, E. D., Ingold, Sir C. K. & Pearson, R. B. (1958). *J. chem. Soc.* p. 4357.
    Blackall, E. L., Hughes, E. D., Ingold, Sir C. K. & Pearson, R. B. (1958).
    J. chem. Soc. p. 4366.
15. Ingold, C. K. (a) with Shaw, F. R. (1927). *J. chem. Soc.* p. 2918; (b) with
    Lapworth, A., Rothstein, E. & Ward, D. (1931). *J. chem. Soc.* p. 1959;
    (c) with Smith, M. S. (1938). *J. chem. Soc.* p. 905.
16. Bird, M. L. & Ingold, C. K. (1938). *J. chem. Soc.* p. 918.
17. Coombes, R. G., Moodie, R. B. & Schofield, K. (1968). *J. chem. Soc.* B,
    p. 800.

4 Nitrating systems:
C. Miscellaneous

4.1 INTRODUCTION

It has been noted (§1.2) that the number of known nitrating systems is large and not surprisingly some of these are based on the oxides of nitrogen; however, nitrous and nitric oxides are not of interest in this connection. Dinitrogen trioxide, the anhydride of nitrous acid, is ionised to give nitrosonium ions in sulphuric acid[1] but such solutions are not nitrating agents:

$$N_2O_3 + 3H_2SO_4 \rightleftharpoons 2NO^+ + H_3O^+ + 3HSO_4^-.$$

More interesting is the behaviour of dinitrogen trioxide in the presence of Lewis acids.[2] Thus, with boron trifluoride it forms a 'complex' which is a good nitrosating (diazotising) agent and a weak nitrating agent.[3] The 'complex' is probably nitrosonium tetrafluoroborate,[4] and it is possible that whatever nitration it does effect proceeds *via* nitrosation followed by oxidation, a process like that discussed below (§4.3.3):

$$3N_2O_3 + 8BF_3 \longrightarrow 6(NO^+)(BF_4^-) + B_2O_3.$$

Solutions of dinitrogen tetroxide (the mixed anhydride of nitric and nitrous acids) in sulphuric acid are nitrating agents (§4.3.2), and there is no doubt that the effective reagent is the nitronium ion. Its formation has been demonstrated by Raman spectroscopy[1a] and by cryoscopy:[1b]

$$N_2O_4 + 3H_2SO_4 \rightleftharpoons NO_2^+ + NO^+ + H_3O^+ + 3HSO_4^-.$$

Nitration has also been effected with the complexes from dinitrogen tetroxide and Lewis acids;[2] in the case of boron trifluoride the complex appears to be a mixture of nitronium and nitrosonium tetrafluoroborates.[4]

Dinitrogen pentoxide being the anhydride of nitric acid, is more fully treated below, as are some other systems with which mechanistic studies have been made.

4.2 NITRATION WITH DINITROGEN PENTOXIDE

4.2.1 *The state of dinitrogen pentoxide in various solvents*

Solid covalent dinitrogen pentoxide can be prepared by freezing the vapour with liquid helium.[5] Normally, solid dinitrogen pentoxide exists as (NO_2^+) (NO_3^-), showing absorption bands in its Raman spectrum[6] only at 1050 and 1400 cm^{-1}; the structure of this form has been determined by X-ray crystallography.[7]

Solutions of dinitrogen pentoxide in nitric acid[8] or sulphuric acid[1a,9] exhibit absorptions in the Raman spectrum at 1050 and 1400 cm^{-1} with intensities proportional to the stoichiometric concentration of dinitrogen pentoxide, showing that in these media the ionization of dinitrogen pentoxide is complete. Concentrated solutions in water (mole fraction of $N_2O_5 > 0.5$) show some ionization to nitrate and nitronium ion.[9]

Dinitrogen pentoxide is not ionized in solutions in carbon tetrachloride, chloroform or nitromethane.[10]

4.2.2 *Nitration in the presence of strong acids or Lewis acids*

Solutions of dinitrogen pentoxide in sulphuric acid nitrate 1,3-dimethylbenzene-4,6-disulphonic acid twice as fast as a solution of the same molar concentration of nitric acid.[11] This is consistent with Raman spectroscopic and cryoscopic data,[1] which establish the following ionisation:

$$N_2O_5 + 3H_2SO_4 \rightleftharpoons 2NO_2^+ + 3HSO_4^- + H_3O^+.$$

When mixed with Lewis acids, dinitrogen pentoxide yields crystalline white solids, which were identified as the corresponding nitronium salts by their infra-red spectra. The reaction with boron trifluoride can be formulated in the following way:[4]

$$3N_2O_5 + 8BF_3 \longrightarrow 6(NO_2^+)(BF_4^-) + B_2O_3.$$

Nitration with pre-formed nitronium salts is discussed below (§4.4).

4.2.3 *Nitration in organic solvents*

Solutions of dinitrogen pentoxide have been used in preparative nitrations.[12a,b] Benzene, bromobenzene, and toluene were nitrated rapidly in solutions of the pentoxide in carbon tetrachloride;[12c] nitrobenzene could not be nitrated under similar conditions, but reacted violently with solid dinitrogen pentoxide.[12c]

Nitrating systems, C

The nitration of sensitive compounds with dinitrogen pentoxide has the advantage of avoiding the use of strong acids or aqueous conditions; this has been exploited in the nitration of benzylidyne trichloride[13a] and benzoyl chloride,[13b] which reacted in carbon tetrachloride smoothly and without hydrolysis.

Kinetic studies of nitration using dilute solutions of dinitrogen pentoxide in organic solvents, chiefly carbon tetrachloride, have provided evidence for the operation, under certain circumstances of the molecular species as the electrophile.[14] The reactions of benzene and toluene were inconveniently fast, and therefore a series of halogenobenzenes and aromatic esters was examined.

The kinetics of the reactions were complicated, but three broad categories were distinguished: in some cases the rate of reaction followed an exponential course corresponding to a first-order form; in others the rate of reaction seemed to be constant until it terminated abruptly when the aromatic had been consumed; yet others were susceptible to autocatalysis of varying intensities. It was realised that the second category of reactions, which apparently accorded to a zeroth-order rate, arose from the superimposition of the two limiting kinetic forms, for all degrees of transition between these forms could be observed.

The observation of two limiting kinetic forms was considered to be symptomatic of the occurrence of two reactions, designated non-catalytic and catalytic respectively. The non-catalytic reaction was favoured at higher temperatures and with lower concentrations of dinitrogen pentoxide, whereas the use of lower temperatures or higher concentrations of dinitrogen pentoxide, or the introduction of nitric acid or sulphuric acid, brought about autocatalysis.

The non-catalytic reaction became dominant in a series of experiments carried out at 10–20 °C, with a concentration of dinitrogen pentoxide in the range 0·03–0·1 mol l^{-1} and containing a large excess of aromatic compound. Under these conditions the reaction obeyed the following kinetic law:

$$\text{rate} = k\,[\text{ArH}]\,[\text{N}_2\text{O}_5].$$

This reaction showed certain characteristics which distinguish it from nitrations in solutions of nitric acid in organic solvents. Thus, in changing the solvent from carbon tetrachloride to nitromethane, the rate increased by a factor of only 6, whereas nitration involving the nitronium ion was accelerated by a factor of about 30 when the solvent was changed from acetic acid to nitromethane. It was held that the insensitivity of the rate to the change in the solvent in the former case was inconsistent with the reaction's being ionogenic. Small concentrations of salts markedly catalysed the reaction, and in this respect nitrate behaved no differently from other salts.

At relatively low temperatures, the effect of added nitric acid was to catalyse the reaction strongly, and to modify it to the autocatalytic form. At higher temperatures the effect of this additive was much weaker, as was the induced autocatalysis. Under these circumstances the catalysis was second-order in the concentration of nitric acid, and the presence of 0.25 mol l^{-1} of it brought about a sixfold change in the rate.

The effect of temperature on the non-catalysed reaction was difficult to disentangle, for at lower temperatures the autocatalytic reaction intervened. However, from a limited range of results, the reaction appeared to have an experimental activation energy of $c. +71$ kJ mol^{-1}.

The catalytic reaction had no simple kinetic form, and its nature depended on many variables; for these reasons the effects of changing the variables can only be discussed qualitatively.

Reducing the temperature or increasing the concentration of reactants, particularly of dinitrogen pentoxide, advanced the onset and increased the intensity of the autocatalysis. Added nitric acid and, to a greater extent sulphuric acid, made the effect more prominent.

The catalysed reaction was considered to arise from the heterolysis of dinitrogen pentoxide induced by aggregates of molecules of nitric acid, to yield nitronium ions and nitrate ions. The reaction is autocatalytic because water produced in the nitration reacts with the pentoxide to form nitric acid. This explanation of the mechanism is supported by the fact that carbon tetrachloride is not a polar solvent, and in it molecules of nitric acid may form clusters rather than be solvated by the solvent (§2.2). The observation that increasing the temperature, which will tend to break up the clusters, diminishes the importance of the catalysed reaction relative to that of the uncatalysed one is also consistent with this explanation. The effect of temperature is reminiscent of the corresponding effect on nitration in solutions of nitric acid in carbon tetrachloride (§3.2) in which, for the same reason, an increase in the temperature decreases the rate.

In the uncatalysed reaction the fact that added nitrate strongly accelerates the rate to the same extent as other salts makes it improbable that the nitronium ion is the effective electrophile. The authors of the work conclude that covalent dinitrogen pentoxide is the electrophilic species, and that added salts exercise their effect by aiding the heterolysis of this molecule so that the nitronium ion becomes the nitrating agent. It was supposed that the insensitivity of the rate to changes in the solvent supported their hypothesis. However, the facts that carbon tetrachloride is a faster solvent than nitromethane in nitration in solutions of nitric acid, and that the addition of acetonitrile or *N,N*-dimethylformamide considerably retards *N*-nitration in solutions of nitric acid in carbon tetrachloride,[15] suggest that the usual considerations of solvent polarity break down when aggregates of molecules are involved.

Further evidence that the nitronium ion was not the electrophile in the uncatalysed reaction, and yet became effective in the catalysed reaction, came from differences in the orientation of substitution. The nitration of chlorobenzene in the uncatalysed reaction yielded only 43 % of the *para* compound, whereas, when the catalysed reaction was made important by adding some nitric acid, the ratio of substitution was that usually observed in nitration involving the nitronium ion (§5.3.4). In the case of the uncatalysed reaction however, the reaction was complicated by the formation of nitrophenols.

4.3 NITRATION VIA NITROSATION

We are not concerned here with the mechanism of nitrosation, but with the anticatalytic effect of nitrous acid upon nitration, and with the way in which this is superseded with very reactive compounds by an indirect mechanism for nitration. The term 'nitrous acid' indicates all the species in a solution which, after dilution with water, can be estimated as nitrous acid.

4.3.1 *The state of nitrous acid in various solvents*

In aqueous solutions of sulphuric (< 50%) and perchloric acid (< 45%) nitrous acid is present predominantly in the molecular form,[16] although some dehydration to dinitrogen trioxide does occur.[17] In solutions containing more than 60% and 65% of perchloric and sulphuric acid respectively, the stoichiometric concentration of nitrous acid is present entirely as the nitrosonium ion[16, 17] (see the discussion of dinitrogen trioxide; §4.1). Evidence for the formation of this ion comes from the occurrence of an absorption band in the Raman spectrum almost identical with the relevant absorption observed in crystalline nitrosonium perchlorate.[18] Under conditions in which molecular nitrous acid and the nitrosonium ion coexist, the sum of the concentrations of these two species accounts for nearly all of the stoichiometric concentration, showing that even if the nitrous acidium ion is formed its concentration is fairly small.[16] There is a change in the ultraviolet spectrum of the nitrosonium ion in sulphuric acid as the concentration is increased above 75 %,[16] and some evidence has been claimed for the existence of an isosbestic point in this change.[19] If this were the case then the existence of a chemical equilibrium would be involved; however, this change is not accompanied by a corresponding change in the Raman spectrum, and it has been suggested that the former change is due to a close association of the nitrosonium ion with the solvent. A similar variation in the ultraviolet spectrum of the nitronium ion has already been remarked upon (§2.4.1).

In an excess of nitric acid, nitrous acid exists essentially as dinitrogen tetroxide which, in anhydrous nitric acid, is almost completely ionised. This is shown by measurements of electrical conductivity,[20] and Raman and infra-red spectroscopy identify the ionic species:[21]

$$N_2O_4 \rightleftharpoons NO^+ + NO_3^-.$$

In mixtures of nitric acid and organic solvents, nitrous acid exists mainly as un-ionised dinitrogen tetroxide.[20] The heterolysis of dinitrogen tetroxide is thus complete in sulphuric acid (§4.1), considerable in nitric acid, and very small in organic solvents.

The condition of dinitrogen trioxide in acid solution is discussed in §4.1.

4.3.2 *The anticatalytic effect of nitrous acid in nitration*

The effect of nitrous acid was first observed for zeroth-order nitrations in nitromethane (§3.2).[22] The effect was a true negative catalysis; the kinetic order was not affected, and nitrous acid was neither consumed nor produced by the nitration. The same was true for nitration in acetic acid.[23] In the zeroth-order nitrations the rate depended on the reciprocal of the square root of the concentration of nitrous acid; $K_{obs} = (a + b\,[HNO_2]_{stoich}^{0.5})^{-1}$. First-order nitrations in the organic solvents follow a law of anticatalysis of the same form (but with different constants in the above equation). With both zeroth- and first-order nitrations a more powerful type of anticatalysis set in when higher (> 0.1 mol l^{-1}) concentrations of nitrous acid were present.

Nitrating systems, C

For nitrations carried out in nitric acid, the anticatalytic influence of nitrous acid was also demonstrated. The effect was smaller, and consequently its kinetic form was not established with certainty. Further, the more powerful type of anticatalysis did not appear at higher concentrations (up to 0·23 mol l^{-1}) of nitrous acid. The addition of water (up to ∼ 5% by volume) greatly reduced the range of concentration of nitrous acid which anticatalysed nitration in a manner resembling that required by the inverse square-root law, and more quickly introduced the more powerful type of anticatalytic effect.

If we consider the effect of nitrous acid upon zeroth-order nitration in organic solvents we must bear in mind that in these circumstances dinitrogen tetroxide is not much ionised, so the measured concentration of nitrous acid gives to a close approximation the concentration of dinitrogen tetroxide. Further, the negligible self-ionisation of nitric acid ensures that the total concentration of nitrate ions is effectively that formed from dinitrogen tetroxide. Consequently as we can see from the equation for the ionisation of dinitrogen tetroxide (§4.3.1),

$$[NO_3^-] \propto [N_2O_4]^{0\cdot5}, \text{ and so } [NO_3^-] \propto [HNO_2]_{stoich.}^{0\cdot5}$$

Now nitrate ions reduce the rate of formation of nitronium ion by deprotonating nitric acidium ions, and this effect must also depend upon $[HNO_2]_{stoich}^{0\cdot5}$, as was observed.

In first-order nitration the anticatalysis is of the same form because the deprotonation of nitric acidium ion diminishes the stationary concentration of nitronium ion and therefore diminishes the rate of nitration.

The weak effect of nitrous acid upon nitration in nitric acid is a consequence of the already considerable concentration of nitrate ions supplied in this case by the medium.

The more powerful anticatalysis of nitration which is found with high concentrations of nitrous acid, and with all concentrations when water is present, is attributed to the formation of dinitrogen trioxide.[23] Heterolysis of dinitrogen trioxide could give nitrosonium and nitrite ions:

$$2N_2O_4 + H_2O \rightleftharpoons N_2O_3 + 2HNO_3.$$
$$N_2O_3 \rightleftharpoons NO^+ + NO_2^-.$$

The anticatalytic action is ascribed to the deprotonation of nitric acidium ions by nitrite ions, which, being more basic than nitrate ions, will be more effective anticatalysts. The effect of nitrite ions should depend upon $[HNO_2]_{stoich}^{1\cdot5}$, as it does.

Anticatalysis by nitrous acid does not occur in the presence of sulphuric acid. In fact, as we have seen (§4.1), dinitrogen tetroxide in concentrated sulphuric acid has been used as a nitrating agent.[24]

4.3.3 *The catalysis of nitration by nitrous acid*

In contrast to its effect upon the general mechanism of nitration by the nitronium ion, nitrous acid catalyses the nitration of phenol, aniline, and related compounds.[11] Some of these compounds are oxidised under the conditions of reaction and the consequent formation of more nitrous acids leads to autocatalysis.

The catalysed nitration of phenol gives chiefly *o*- and *p*-nitrophenol, ($< 0.1\%$ of *m*-nitrophenol is formed),[26a] with small quantities of dinitrated compound and condensed products. The *ortho:para* ratio is very dependent on the conditions of reaction and the concentration of nitrous acid.[26b] Thus, in aqueous solution containing sulphuric acid (1.75 mol l^{-1}) and nitric acid (0.5 mol l^{-1}), the proportion of *ortho*-substitution decreases from 73 % to 9 % as the concentration of nitrous acid is varied from 0–1 mol l^{-1}. However, when acetic acid is the solvent the proportion of *ortho*-substitution changes from 44 % to 74 % on the introduction of dinitrogen tetroxide (4.5 mol l^{-1}).

The theory that the catalysed nitration proceeds through nitrosation was supported by the isolation of some *p*-nitrosophenol from the interrupted nitration of phenol,[26c] and from the observation that the *ortho:para* ratio ($9:91$) of strongly catalysed nitration under aqueous conditions was very similar to the corresponding ratio of formation of nitrosophenols in the absence of nitric acid.[27]

The nitration of anisole in 40 % aq. nitric acid in the presence of some nitrous acid yielded 2,4-dinitrophenol as the main product.[25c] In more concentrated solutions of nitric acid *o*- and *p*-nitroanisoles were the main products,[25c] less than 0.1 % of the *meta*-isomer being formed.[26a] The isomeric ratios for nitration under a variety of conditions are given later (§5.3.4).

The kinetics of nitration of anisole in solutions of nitric acid in acetic acid were complicated, for both autocatalysis and autoretardation could be observed under suitable conditions.[26b] However, it was concluded from these results that two mechanisms of nitration were operating, namely the general mechanism involving the nitronium ion and the reaction catalysed by nitrous acid. It was not possible to isolate these mechanisms completely, although by varying the conditions either could be made dominant.

Nitrating systems, C

p-Chloroanisole and p-nitrophenol, the nitrations of which are susceptible to positive catalysis by nitrous acid, but from which the products are not prone to the oxidation which leads to autocatalysis, were the subjects of a more detailed investigation.[26] With high concentrations of nitric acid and low concentrations of nitrous acid in acetic acid, p-chloroanisole underwent nitration according to a zeroth-order rate law. The rate was repressed by the addition of a small concentration of nitrous acid according to the usual law: rate $= k_0(1 + a[HNO_2]^{0.5}_{stoich})^{-1}$. The nitration of p-nitrophenol under comparable conditions did not accord to a simple kinetic law, but nitrous acid was shown to anticatalyse the reaction.

By using higher concentrations of nitrous acid, and reducing that of nitric acid, the nitration of both compounds was brought under the control of the following rate law:

$$rate = k[ArH][HNO_2]_{stoich}.$$

The catalysis was very strong, for in the absence of nitrous acid nitration was very slow. The rate of the catalysed reaction increased steeply with the concentration of nitric acid, but not as steeply as the zeroth-order rate of nitration, for at high acidities the general nitronium ion mechanism of nitration intervened.

The effect of nitrous acid on the nitration of mesitylene in acetic acid was also investigated.[26b] In solutions containing 5–7 mol l^{-1} of nitric acid and $< c.$ 0·014 mol l^{-1} of nitrous acid, the rate was independent of the concentration of the aromatic. As the concentration of nitrous acid was increased, the catalysed reaction intervened, and superimposed a first-order reaction on the zeroth-order one. The catalysed reaction could not be made sufficiently dominant to impose a truly first-order rate. Because the kinetic order was intermediate the importance of the catalysed reaction was gauged by following initial rates, and it was shown that in a solution containing 5·7 mol l^{-1} of nitric acid and 0·5 mol l^{-1} of nitrous acid, the catalysed reaction was initially twice as important as the general nitronium ion mechanism.

The observation of nitration via nitrosation for mesitylene is important, for it shows that this reaction depends on the reactivity of the aromatic nucleus rather than on any special properties of phenols or anilines.

The mechanism for the reaction can be represented by the following equations:

$$ArH + HNO_2 \xrightarrow{slow} ArNO + H_2O,$$
$$ArNO + HNO_3 \xrightarrow{fast} ArNO_2 + HNO_2.$$

This scheme accounts for the facts that the concentration of nitrous acid does not change during reaction, and that nitro compounds rather than nitroso compounds are the predominant products even when the reaction is interrupted before completion. However, it is a feature of nitrosation that many species are known to be effective agents;[28] in the present case the kinetics are consistent with the operation of NO^+, $H_2NO_2^+$, HNO_2, and N_2O_4, acting individually or together. In the solutions used, analytical nitrous acid is present mainly as dinitrogen tetroxide, although this is partially ionised to the nitrosonium and nitrate ions. The concentration of nitrate ions arising from the ionisation of nitric acid will be greater than that produced in the above equilibrium, and therefore the concentration of nitrosonium ions will bear a constant ratio to that of its precursor.

Added nitrate ions, and to a smaller extent water, depress the rate of the catalysed reaction,[29] therefore excluding the operation of HNO_2 and $H_2NO_2^+$. The depression of the rate by nitrate obeys the following expression:

$$k_{obs} = a + b \,[NO_3^-]^{-1}.$$

This indicates that both N_2O_4 and NO^+ are involved in the reaction

$$N_2O_4 + ArH \xrightarrow{\ k_1\ } ArNO + HNO_3,$$

$$NO^+ + ArH \xrightarrow{\ k_2\ } ArNO + H^+,$$

and because dinitrogen tetroxide is the predominant species, the rate can be written in the following way, where K is the equilibrium constant for the formation of the nitrosonium ion ($N_2O_4 \rightleftharpoons NO^+ + NO_3^-$):

$$k_{obs} = k_1 + k_2 K[NO_3^-]^{-1}.$$

It was estimated from an analysis of the results that the nitrosonium ion was at least ten times more effective than dinitrogen tetroxide; this is a lower limit, and the ion is likely to be much more reactive than the latter species.

It has been considered that nitric acid was responsible for the oxidation of the nitroso compound, but there is recent evidence from the catalysed nitration of p-dimethoxybenzene in carbon tetrachloride that dinitrogen tetroxide is involved:[30]

$$N_2O_4 + ArNO \longrightarrow ArNO_2 + N_2O_3,$$

$$N_2O_3 + 2HNO_3 \longrightarrow 2N_2O_4 + H_2O.$$

4.3.4 *Nitration at the encounter rate and nitrosation*

As has been seen (§3.3), the rate of nitration by solutions of nitric acid in nitromethane or sulpholan reaches a limit for activated compounds which is about 300 times the rate for benzene under the same conditions. Under the conditions of first-order nitration (7·5 % aqueous sulpholan) mesitylene reacts at this limiting rate, and its nitration is not subject to catalysis by nitrous acid; thus, mesitylene is nitrated by nitronium ions at the encounter rate, and under these conditions is not subject to nitration *via* nitrosation. The significance of nitration at the encounter rate for mechanistic studies has been discussed (§2.5).

Under the conditions mentioned, 1-methylnaphthalene was nitrated appreciably faster than was mesitylene, and the nitration was strongly catalysed by nitrous acid. The mere fact of reaction at a rate greater than the encounter rate demonstrates the incursion of a new mechanism of nitration, and its characteristics identify it as nitration *via* nitrosation.

Under the same conditions the even more reactive compounds 1,6-dimethylnaphthalene, phenol, and *m*-cresol were nitrated very rapidly by an autocatalytic process [nitrous acid being generated in the way already discussed (§4.3.3)]. However, by adding urea to the solutions the autocatalytic reaction could be suppressed, and 1,6-dimethyl-naphthalene and phenol were found to be nitrated about 700 times faster than benzene. Again, the barrier of the encounter rate of reaction with nitronium ions was broken, and the occurrence of nitration by the special mechanism, *via* nitrosation, demonstrated.

In 7·5 % aqueous sulpholan the very reactive compound, anthanthrene, could be nitrated according to a first-order law only with low concentrations of nitric acid, and the reaction was very strongly catalysed by nitrous acid. Under zeroth-order conditions (i.e. in the absence of water and with $[HNO_3] = 5$ mol l^{-1}) and with a very small concentration of nitrous acid ($[HNO_2] < 3 \times 10^{-5}$ mol l^{-1}; $[urea] = 0·05$ mol l^{-1}), where the use of mesitylene gave $k_0 = 2·1 \times 10^{-6}$ mol l^{-1} s^{-1} (at 25 °C), the nitration of anthanthrene was too fast to be measured. Clearly, the nitronium ion mechanism could not be operative. With low concentrations of nitric acid ($[HNO_3] < 1$ mol l^{-1}) zeroth-order nitration of anthanthrene of the same rate as that for mesitylene could with difficulty be observed; often autocatalysis intervened.[31]

4.4 NITRATIONS WITH SOLUTIONS OF NITRONIUM SALTS IN ORGANIC SOLVENTS

4.4.1 *Preparation and properties of nitronium salts*

The first preparation of a nitronium salt by Hantzsch, who isolated the perchlorate mixed with hydroxonium perchlorate, and some of the subsequent history of these salts has already been recounted (§2.3.1).

Nitronium tetrafluoroborate was first prepared by adding a mixture of anhydrous hydrofluoric acid and boron trifluoride to a solution of dinitrogen pentoxide in nitromethane.[32] Nitric acid can be used in place of dinitrogen pentoxide, and by replacing boron trifluoride by other Lewis-acid fluorides Olah and his co-workers prepared an extensive series of stable nitronium salts.[2]

Nitronium salts are colourless, crystalline and very hygroscopic; nitronium perchlorate and sulphate are unstable and liable to spontaneous decomposition, whereas nitronium tetrafluoroborate and other complex fluoro-salts are relatively stable.

Cryoscopic investigations suggest that in sulpholan nitronium tetrafluoroborate exists predominantly as ion pairs.[2, 33d] The specific conductivity of these solutions increases linearly with the concentration of the salt (up to 0·4 mol l^{-1}),[34] and is attributed to the existence of iontriplets rather than free ions.[33d]

4.4.2 *The use of nitronium salts in nitration*

Olah's original preparative nitrations were carried out with mixtures of the aromatic compound and nitronium salt alone or in ether,[33a] and later with sulpholan as the solvent.[34] High yields of nitro-compounds were obtained from a wide range of aromatic compounds, and the anhydrous conditions have obvious advantages when functional groups such as cyano, alkoxycarbonyl, or halogenocarbonyl are present.[34] The presence of basic functions raises difficulties; with pyridine no C-nitration occurs, 1-nitropyridinium being formed.[35]

The selection of solvents for quantitative work is not easy. Nitro-alkanes are sufficiently inert, but nitronium tetrafluoroborate is poorly soluble in them (*c.* 0·3%). Nitronium salts react rapidly with acetic anhydride, and less rapidly with acetic acid, *N,N*-dimethylformamide and acetonitrile, although the latter solvent can be used for nitration at low temperatures. Sulpholan was selected as the most suitable solvent;

it is inert to these salts, and a solution in it of 0·5 mol l⁻¹ of nitronium tetrafluoroborate can be prepared.[34]

Quantitative comparisons of aromatic reactivities were made by using the competitive method with solutions of nitronium tetrafluoroborate in sulpholan, and a concentration of aromatic compounds 10 times that of the salt. To achieve this condition considerable proportions of the aromatic compounds were added to the medium, thus depriving the sulpholan of its role as true solvent; thus, in the nitration of the alkyl-[33b] and halogeno-benzenes,[33c] the description of the experimental method shows that about 50–60 cm³ of mixed aromatic compounds were dissolved in a total of 130 cm³ of sulpholan.

We are not, at this stage, primarily concerned with the results of using nitration processes as a means of comparing aromatic reactivities, but the results obtained using nitronium salts in organic solvents must be discussed to some extent now, because of questions they raise about nitration processes. The results obtained differ in several respects from those from other methods of nitration. The most important change is that differences in reactivity from one compound to another are very much decreased. The diminution of inter-molecular selectivity is not accompanied by loss of intra-molecular selectivity; the proportions of isomers formed do not differ grossly from those obtained from other methods. Thus, toluene, which other methods of nitration show to be 17–28 times more reactive than benzene appears from Olah's results to be less than twice as reactive. These and results from other alkyl-benzenes are collected in tables 4.1 and 4.2. The relative rates tended towards unity with rising temperature,[33b,c] hexadeuterobenzene reacted about 12% faster than benzene,[33b] and the relative rates varied somewhat with the solvent and with the anion associated with the nitronium ion[33e]. Two surprising consequences are that the nitration of toluene with nitronium tetrafluoroborate indicates the *meta* position to be deactivated (table 4.2), and that nitration of mesitylene with nitronium hexafluorophosphate or hexafluoroarsenate in nitromethane suggests that a position in mesitylene is less reactive than one in benzene (relative rates: 0·41, and 0·42, respectively).

As a means of determining relative reactivities, the competition method using nitronium tetrafluoroborate in sulpholan has been criticised as giving results which arise from incomplete mixing of the reagents before reaction is complete. The difficulty of using the competition method when the rate of reaction is similar to, or greater than,

TABLE 4.1 *Nitration of aromatic compounds: relative rates at 25 °C*

Compound	(a)	(b)	(c)	(d)	(e)	(f)	(g)	(h)	(i)	(j)	(k)	(l)	(m)	(n)	(o)	(p)	(q)
Benzene	1	1	1	1	1	1	1	1	1	1	1	1	1	1	1	1	1
Toluene	1·67	23	26·4	28·8	17	28	21	1·60	2·13	1·24	17	19	20	25	37		28
Ethylbenzene	1·60	23	22·6			24		1·35								1·65	
n-Propylbenzene	1·46																
iso-Propylbenzene	1·32	18				13·8											
n-Butylbenzene	1·39																
tert-Butylbenzene	1·18	15															
o-Xylene	1·75		$>10^3$	$>10^3$				0·9		1·02	38		61	139			
m-Xylene	1·65		$>10^3$	$>10^3$				1·1	1·27	0·80	38		100	146			
p-Xylene	1·96		$>10^3$	$>10^3$		>500	>500	1·8		1·09	38	84	114	130			
Mesitylene	2·71		$>10^3$	$>10^3$	$>10^3$	$>10^3$	$>10^3$	0·33		0·68	36	78	350	400			
Fluorobenzene	0·45	0·15									0·117[51]	0·17[51]					
Chlorobenzene	0·14	0·033									0·064[51]	0·065[51]					
Bromobenzene	0·12	0·030									0·066[51]	0·065[51]					
Iodobenzene	0·28	0·18									0·125[51]	0·28[51]					

(a) Nitronium tetrafluoroborate in sulpholan.[33b,c] (b) For toluene, acetyl nitrate in acetic anhydride at 30 °C.[46a] For the other cases a reagent prepared from fuming nitric acid and acetic anhydride,[47] or from nitric acid and acetic anhydride.[33d]. For the halogenobenzenes, acetyl nitrate in acetic anhydride at 18 °C.[46b] See also ref. 48. (c) Nitric acid in nitromethane.[33d,46a] (d) Nitric acid in acetic acid.[33d] (e) Nitric acid in sulpholan.[33d] See ref. 31 for comments on the composition of the medium. (f) 30% solution of mixed acid (1:1) in sulpholan.[33d] (g) 30% solution of mixed acid (1:1) in acetic acid.[33d] (h) 75% solution of mixed acid (1:1) in sulpholan.[33d] (i) 75% solution of mixed acid (1:1) in acetic acid.[33d] (j) Heterogeneous nitration with 1:1 mixed acid.[33d] (k) Nitric acid ($\sim 9 \times 10^{-4}$ mol l^{-1}) in 68·3% sulphuric acid ($[ArH]$ = c. 10^{-4}–10^{-5} mol l^{-1}).[49] (l) Similar to (k) but in 61·05% perchloric acid.[49] (m) Nitric acid in 75% aqueous sulpholan.[31] (n) Nitric acid in 15% aqueous sulpholan.[31] (o) Nitric acid (1 mol l^{-1}) in sulpholan.[40] (p) 100% nitric acid.[40] (q) Nitric acid in trifluoroacetic acid.[50]

TABLE 4.2 *Nitration of aromatic compounds: isomer proportions and partial rate factors**

Compound	Nitrating system (table 4.1)	ortho	meta	para	$\frac{1}{2}o:p$- ratio	f_o	f_m	f_p
		Isomer proportions (%)				Partial rate factors		
Toluene	(a)	65·4	2·8	31·8	1·03	3·27	0·14	3·18
	(b)	58·1	3·7	38·2	0·76	47·1	3·0	61·9
	(c)	61·5	3·1	35·4	0·87	49	2·5	56
	(d)	56·9	2·8	40·3	0·71	49	2·4	70
	(e)	61·9	3·5	34·7	0·89	32	1·7	35
	(f)	62·0	3·4	34·6	0·89	52·1	2·8	58·1
	(g)	56·5	3·1	40·4	0·70	·	·	·
	(h)	56·3	2·6	41·0	0·69	·	·	·
	(i)	58·1	1·9	40·0	0·73	·	·	·
	(j)	56·4	4·8	38·8	0·73	·	·	·
	(k)	60±5	3±1	37±5				
	(o)	64·9	5·6	29·5	1·10	·	·	·
	(p)	57·5	4·6	37·9	0·76	·	·	·
	(q)	61·6	2·6	35·8	0·86	51·7	2·18	60·1
Ethylbenzene	(a)	53·0	2·9	44·1	0·60	·	·	·
	(b)	45·9	3·3	50·8	0·45	31·4	2·3	69·5
	(c)	48·3	2·3	49·5	0·49	32·7	1·6	67·1
	(f)	50·3	3·6	46·1	0·55	36·2	2·6	66·4
	(h)	44·7	2·0	53·3	0·42			
n-Propylbenzene	(a)	51·0	2·3	46·7	0·55			
iso-Propylbenzene	(a)	23·4	6·9	69·7	0·17	·	·	·
	(b)	28·0	4·5	67·5	0·21	14·8	2·4	71·6
	(f)	43·2	4·5	52·3	0·41	17·9	1·9	43·3
n-Butylbenzene	(a)	50·0	2·0	48·0	0·52	·	·	·
tert-Butylbenzene	(a)	14·3	10·7	75·0	0·095	·	·	·
	(b)	10·0	6·8	83·2	0·06	4·5	3·0	75·5
o-Xylene	(a)	3-NO₂ 79·7; 4-NO₂ 20·3				·	·	·
	(j)	55		45		·	·	·
m-Xylene	(a)	2-NO₂ 17·8; 4-NO₂ 82·2				·	·	·
	(h)	15·3		·		·	·	·
	(i)	16·3		83·8		·	·	·
	(j)	14		86		·	·	·
Fluorobenzene	(a)	8·5	·	91·5		·	·	·
	(b)	·	·	·		·	·	·
	(k)	13	0·6	86		·	·	·
Chlorobenzene	(a)	22·7	0·7	76·6		·	·	·
	(b)	29·6	0·9	69·5		·	·	·
	(k)	33·5	1·1	65·5		·	·	·
Bromobenzene	(a)	25·7	1·1	73·2		·	·	·
	(b)	36·5	1·2	62·4		·	·	·
	(k)	45	0·9	54		·	·	·
Iodobenzene	(a)	36·3	—	63·7		·	·	·
	(b)	38·3	1·8	59·7		·	·	·
	(k)	45	1·3	54		·	·	·

* For related data see tables 5.2, 9.1 and 9.5.

the rate of mixing has long been appreciated.[36] It is interesting to consider what result might be expected from the conditions originally used[33b] (a solution of 0·05 mol of nitronium tetrafluoroborate in 60 g of sulpholan was added drop by drop to a vigorously stirred solution of 0·25 mol of benzene and 0·25 mol of alkylbenzene in 70 g of sulpholan at 25 °C) if this were so. Two processes are involved: the mixing of the two solutions and the reactions of the electrophile with the aromatics. The faster the latter process, the more completely will the observed result be determined by the process of mixing; the faster the former process, the more completely will the observed result be a consequence of the relative speeds of the two nitration steps. Two limiting conditions obviously exist; when the nitration processes are both very fast compared with mixing, products will be formed very closely in the ratio in which the two aromatics are present in the mixture, and the relative rate will be effectively unity; when mixing is very fast compared with nitration, so that solutions are in effect homogeneous, the ratio of products will give a true measure of the relative rates of nitration of benzene and the alkylbenzene. A special case of the first kind would be that in which the two competing aromatics both reacted upon encounter with the nitrating agent; the relative rate would then necessarily be close to unity (§3.3).

In all other cases the observed result will depend upon both the speed of mixing and the speed of nitration. The relative rate will be greater than unity by an amount peculiar to the conditions of the experiment. Again, if the alkylbenzene is sufficiently reactive to be nitrated upon encounter, whilst benzene is not, the relative rate will be greater than unity and, for the experimental conditions, will be a limiting upper value no matter what aromatic is used.

The situation, and the slightly more complicated one arising when the proportion of alkylbenzene to benzene in the original mixture is varied, can be most simply discussed using a model of the kind mentioned by Ridd and his co-workers.[37a] Suppose some solution containing 0·05 mol of nitronium tetrafluoroborate mixes with a solution containing benzene and a homologue, so that all of the nitrating agent is consumed by the time that it has mixed with a volume of solution containing 1·2 equivalents of aromatics (the particular figure is not essential to the argument, but is not entirely arbitrary, as will be seen later). Under the conditions of the original experiments there would be formed 0·03 mol of alkylnitrobenzenes and 0·02 mol of nitrobenzene, giving a relative rate of 1·5,* The results for mixtures containing various proportions of the two aromatics are shown in table 4.3.

TABLE 4.3 *Product ratios arising from nitration under conditions where mixing is slow*

	Molar ratio of alkylbenzene to benzene	No. of equivs. of aromatics in solution in which nitrating agent is consumed	Products		Relative rate
			Alkylnitro-benzenes (mol)	Nitrobenzene (mol)	
(a)	1:1	1·2	0·03	0·02	1·5
(b)	4:1	1·2	0·048	0·002	6·0
(c)	4:1	1·1	0·044	0·006	1·8
(d)	4:1	1·0	0·025	0·025	1·0
(e)	1:4	1·2	0·012	0·038	1·2
(f)	1:4	1·5	0·015	0·035	1·7
(g)	1:4	1·0	0·01	0·04	1·0

(a) is the case just discussed. In (b), although the ratio of alkylbenzene to benzene is now 4:1, it is still supposed that the nitrating reagent traverses a volume of solution containing 1·2 equivalents of aromatics; the relative rate of 6·0 is now an upper limit, for with a greater concentration of the more reactive aromatic the volume of solution traversed would be smaller. It might, as in (c) containing only 1·1 equivalents of aromatics with the consequences shown. In (e) we have the reverse circumstances, and the relative rate of 1·2 then represents a lower limit, for now more solution would be traversed, with the sort of consequence seen in (f). (d) and (g) both show the result of very fast nitration steps, in the limit occurring upon encounter.

Two things are clear; where mixing is a slow process, relative rates can give a very misleading picture, and varying the proportions of aromatics will not necessarily clarify the position. The above model makes it quite understandable that varying the proportions of aromatics could leave the apparent relative rates unchanged. Also, the result need not, and indeed most generally will not, be just a statistical one.

The results of Olah and his co-workers[33, 34] require, then, no special hypothesis regarding mechanism of nitration, in so far as they relate to

---

* 'Relative rate' meaning, here, simply the ratio of nitro-alkylbenzene to nitrobenzene, multiplied by the initial ratio of alkylbenzene to benzene. This is not precisely the same as the ratio of rate constants for nitration.[38]

intermolecular selectivities and the effect upon them of varying the proportions of aromatics in the nitration solution.

In considering the possible influence of the rate of mixing upon their results, Olah, Kuhn, and Flood[33b] examined the effect of changing the rate of stirring; it was stated to be without effect. However, Tolgyesi found that lowering the concentration of the nitronium salt and increasing the rate of mixing in nitrations with nitronium tetrafluoroborate raised the apparent reactivity of toluene compared with that of benzene to that (25–30) found for nitrations with nitric acid.[39] Olah and Overchuk[40b] re-examined these results and discounted them, largely on the grounds that the nitronium salt reacted with impurities in such dilute solutions giving another electrophile. Furthermore, other workers[41] presented results in accord with Olah's original observations and failing to provide evidence for the importance of mixing.

Olah and Overchuk[40b] also attempted to discover evidence of slow mixing by carrying out reactions in high-speed flow systems. Evidence, including the isolation of dinitro compounds ($>$ 1 %), was indeed found, but held to show that the effect of imperfect mixing was only minor. The reactions were, unfortunately, too fast to permit determinations of absolute rates (half-lives of about 10^{-3} s).

More recently, quite unambiguous evidence has been obtained for the view that in nitrations with solutions of nitronium tetrafluoroborate in sulpholan incomplete mixing is an important factor.[37] Dibenzyl (up to a fivefold excess) in sulpholan was nitrated by the addition of a solution of the nitronium salt ($\sim$ 0·2 M). Analysis of the products suggested that the proportion of dinitro compounds decreased with the efficiency of mixing. More important is the actual amount of dinitro compound formed, for if mixing is complete before reaction the relative amounts of nitro- and di-nitro-dibenzyl should be those expected from the random pairing of benzene rings, for nitration of one ring should have a negligible influence upon the reactivity of the other. Thus, with the concentrations of reactants used by Ridd and his co-workers, where one in four of the aromatic rings could be nitrated, the proportions would be $\frac{1}{16}$ of dinitro-dibenzyl, $\frac{6}{16}$ of mononitrodibenzyl and $\frac{9}{16}$ of dibenzyl. About 14 % of the nitrated product would be dinitro-dibenzyl. In fact, the observed proportion was 54–77 %. Clearly, mixing was slow by comparison with nitration, and some molecules had a greater chance of reaction than had others. Further analysis of the results suggested that the nitronium ion was reactive enough to cause complete reaction in the time needed for mixing to have occurred with about 1·2 equivalents of the dibenzyl (see above).

The reactivity of dibenzyl is similar to those of alkylbenzenes, and it is therefore most probable that the nitrations of the latter substances were also influenced by mixing.*

The data of table 4.1, column (*a*), do not suggest that nitration of the alkylbenzenes with nitronium tetrafluoroborate in sulpholan occurs upon encounter; mesitylene might be so reacting (§3.3) but comparison with a more inherently reactive substance would be necessary before this possibility could be considered.

It can be concluded, as already stated above, that the diminution in intermolecular selectivity observed in these nitrations with nitronium salts in organic solvents does not of itself require any special mechanistic considerations as regards the process of substitution.

It has already been noted that, as well as alkylbenzenes, a wide range of other aromatic compounds has been nitrated with nitronium salts. In particular the case of nitrobenzene has been examined kinetically.[42] Results are collected in table 4.4. The reaction was kinetically of the first order in the concentration of the aromatic and of the nitronium salt. There is agreement between the results for those cases in which the solvent induces the ionization of nitric acid to nitronium ion, and the corresponding results for solutions of preformed nitronium salts in the same solvent.

In nitration with nitronium salts in sulpholan, nitrobenzene was substituted in the following proportions: 8% *ortho*, 90% *meta* and 2% *para*;[40a] under the same conditions benzylidyne trifluoride yielded 8%, 88% and 4% of *o*-, *m*- and *p*-nitro compound respectively.[40a] Both of these aromatic compounds were stated to be 10^{-3}–10^{-4} times less reactive than benzene.[40a]

4.4.3 *Comparisons with other systems*

Olah and his co-workers[33d] compared the behaviour of nitronium salts in competitive nitrations with the behaviour of other nitrating systems. The results are given in table 4.1, columns (*a*)–(*j*), and also in table 4.2. The results obtained from competitive nitrations using solutions of nitric acid in organic solvents (table 4.1, columns (*b*)–(*e*)) are in line with those obtained by earlier workers. The evidence that in nitromethane,

---

* Work on the nitration of durene also gives evidence for the importance of the speed of mixing in nitration with nitronium salts.[52]

TABLE 4.4   *The kinetics of nitration of nitrobenzene in various media*[42]

Solvent	Temp./° C	Nitrating agent	$\dfrac{k_2}{\text{l mol}^{-1}\,\text{s}^{-1}}$
Methanesulphonic acid	25	$NO_2{}^+BF_4{}^-$	$1\cdot83 \times 10^{-2}$
Sulphuric acid*	23	$NO_2{}^+BF_4{}^-$	$7\cdot3 \times 10^{-3}$
Sulphuric acid†	23	$NO_2{}^+BF_4{}^-$	$5\cdot3 \times 10^{-3}$
Acetonitrile	21	$NO_2{}^+BF_4{}^-$	$1\cdot8 \times 10^{-3}$
Methanesulphonic acid	~25	HNO_3	$\sim2 \times 10^{-2}$
Sulphuric acid*	~25	HNO_3	$7\cdot0 \times 10^{-2}$
Nitric acid	-13	.	$\sim2 \times 10^{-2}$
Nitric acid	21	.	$\sim1\cdot7 \times 10^{-1}$
Acetonitrile	~21	HNO_3	$<2 \times 10^{-5}$

* $[H_2O] = c.\ 0\cdot12$ mol l^{-1}.   † $[SO_3] = c.\ 0\cdot18$ mol l^{-1}.

acetic acid and sulpholan the nitronium ion is the electrophile has been presented (§3.2). The high relative reactivities ($> 10^3$) reported for *p*-xylene and mesitylene are not consistent with the establishment of a limiting rate of nitration upon encounter in these solvents[31] (§3.3). The problem of nitration in acetic anhydride is discussed in §5.3.

In the experiments using a 1:1 mixture of nitric and sulphuric acids (table 4.1, column (*j*)) reaction occurred under heterogeneous conditions, about 50 cm³ of mixed aromatic compounds and 25 cm³ of mixed acids being used. The results are therefore complicated by differences in solubilities and rates of diffusion to the acid layer.

Even so, the results were claimed to show a greater resemblance to nitrations with nitronium salts than to nitrations in organic solvents. However, reaction at the encounter rate (§3.3) imposes a limit to the rate of reaction in these media, which decreases from 40 times the rate for benzene in 68 % sulphuric acid to 6 times the rate in 80 % sulphuric acid. Therefore it is reasonable to expect that in stronger solutions even under homogeneous conditions, the rates of these compounds would approximate to that of benzene.

The use of 1:1 mixed acid in sulpholan and in acetic acid was examined (table 4.1, columns (*f*)–(*i*)). The variation of the concentration of nitronium ions with the concentration of mixed acids ($[H_2SO_4]:[HNO_3]$, 1:1), in sulpholan (*a*), acetic acid (*b*), and nitromethane (*c*) are illustrated in fig. 4.1. The results for acetic acid and sulpholan were determined by Raman spectroscopy, and those for nitromethane from the infra-red spectra.

69

Nitrating systems, C

Using sulpholan and acetic acid as solvents competitive nitrations were performed with solutions containing 75% and 30% of mixed acid (table 4.1, columns h, i and f, g, respectively). In the former the concentration of nitronium ions was substantial (c. 5–7% by weight), whereas in the latter the concentration was below the level of spectroscopic detection.

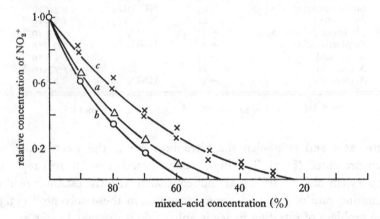

Fig. 4.1. Variation of NO_2^+ ion concentration with the concentration of mixed acid (nitric:sulphuric, 1 mole:1 mole) in organic solvents: (a) in sulpholan; (b) in acetic acid; (c) in nitromethane. Curves (a) and (b) were determined by Raman measurements using the 1400 cm⁻¹ band while curve (c) was derived from infra-red measurements on the 2375 cm⁻¹ band. Unity on the NO_2^+ concentration scale was determined to be 5·6 molar ($\sim$25·8 weight %). (From Olah et al.[43])

With the more concentrated solution the results, as regards loss of intermolecular selectivity, were similar to those obtained with nitronium salts (table 4.1, column a), whilst with the more dilute solution a more usual situation was revealed. The significance of the former observations is again open to doubt because of the likelihood that mixing was relatively slow, and also because reaction upon encounter is here a serious probability.

It has been necessary to comment upon these various studies because Olah and his co-workers[33b, d] have suggested that whilst nitrations, like those with nitronium salts, which give a relative rate of reaction of toluene with respect to benzene not much greater than unity involve the nitronium ion as the electrophile, this is not so in other cases. It is important to consider these opinions closely. In the earlier of the two relevant papers[33b] it is agreed that since nitrations of toluene with nitronium tetrafluoroborate in sulpholan show no abnormal o:p-ratio there is no abnormal steric effect and therefore 'the nitrating species (NO_2^+) must exist as a loosely solvated or unsolvated ion-pair, if not as free nitronium ions'. Further, the reaction is considered

informative—as a specific example of an electrophilic aromatic substitution, i.e. the interaction of a preprepared, stable cation (NO_2^+) with sufficiently reactive aromatics such as benzene and alkylbenzenes... useful information was obtained about the nature of interaction of the electrophile (NO_2^+) with the aromatic substrates... no generalisation from present observations for other electrophilic aromatic substitution is possible; certainly not for reactions where the formation of the electrophilic reagent involves a slow kinetic step.

In pursuing the point about the differences between the case where the electrophile is preformed and where it is formed in a slow step the authors remark:[33d]

The effect of aromatic substrates on the formation of NO_2^+ is shown in the considerably increased substrate selectivity over that obtained with NO_2^+ salts. On the basis of the experimental data it is suggested that in these nitrations a weaker nitrating species than NO_2^+ must be involved in the primary interaction with the aromatic substrates. This incipient nitronium ion then attaches itself to the aromatics in a step giving high substrate selectivity. Whether the incipient nitronium ion is the nitracidium ion ($H_2NO_3^+$), protonated acetyl nitrate ($CH_3COO—HNO_2^+$) or probably a transition state of any of those unstable species to NO_2^+, in which water is loosened, but not yet completely eliminated, is difficult to say and no direct physical evidence is available.

It is explicitly stated that[33d]:

...the reagent which nitrates more active aromatic molecules at a speed much greater in the same medium* and which is also effective in more aqueous media, still needs to be identified. It may be the nitracidium ion or the nitronium ion present in small concentrations, or any intermediate state between the two limiting cases, the aromatic substrate interacting with the precursor of NO_2^+ before it is entirely preformed.

The clear ideas which seem to emerge from these statements are:

(1) The nitronium ion is the electrophile in nitrations with nitronium salts in organic solvents.

(2) In other systems the nitronium ion is also effective with de-activated substrates (such as nitrobenzene in 80 % sulphuric acid).

(3) In other systems electrophiles other than the nitronium ion are involved with activated substrates (in these cases intermolecular selectivity is high, whereas with nitronium salts it is low).

(4) The nitration of alkylbenzenes and benzene with nitronium salts

* i.e. faster than nitrobenzene in 80% sulphuric acid.

71

in organic solvents provides information about the interaction of the electrophile with the aromatic molecule.

About (1) and (2) there can be no dispute, but (3) must be rejected. The implication that the nitronium ion, effectively freed from a close association with another entity, is not the nitrating agent in those reactions of benzene and its homologues, under conditions in which substantial intermolecular selectivity is observed, conflicts with previous evidence (§3.2). Thus, in nitration in organic solvents and in aqueous nitric acids, the observation of kinetically zeroth-order nitration, and the effect of added nitrate on this rate, is compelling evidence for the operation of the nitronium ion. The nitric acidium ion is not the electrophile under these conditions, and it is difficult to envisage how a species in which 'the water is loosened but not yet completely eliminated' could be formed in a slow step independent of the aromatic and be capable of a separate existence. It is implicit that this species should be appreciably different from the nitronium ion in its electrophilic properties. There is no support to be found for the participation of the aromatic in the formation of the electrophile.

The possibility mentioned in (4) is of great importance, and centres on the persistence in nitrations with nitronium salts of positional selectivities in the nitration of alkylbenzenes. It is discussed in §6.2.

It is worth noting here that the results of some other studies of aromatic substitutions, such as the Friedel–Crafts benzylation[43a] and isopropylation[43c] of alkylbenzenes, and the bromination of alkylbenzenes with bromine, catalysed by ferric chloride,[43b] are under suspicion as depending upon slow mixing. As regards halogenation catalysed by Lewis acids, positive evidence to support this criticism has been obtained.[44]

4.4.4 *Nitrations with 1-nitropyridinium salts*

It has been mentioned (§ 4.4.2) that nitronium tetrafluoroborate reacts with pyridine to give 1-nitropyridinium tetrafluoroborate. This compound and several of its derivatives have been used to effect what is called the 'transfer nitration' of benzene and toluene.[45] 1-Nitropyridinium tetrafluoroborate is only sparingly soluble in acetonitrile, but its homologues are quite soluble and can be used without isolation from the solution in which they are prepared. 1-Nitropyridinium tetrafluoroborate did nitrate toluene in boiling acetonitrile slowly, but not at 25 °. In contrast, 1-nitro-2-picolinium tetrafluoroborate readily

effected quantitative nitration at ordinary temperatures; presumably
the 2-methyl group, by preventing planarity, weakened the bonding
between the nitro group and the pyridine ring. The results of competitive nitrations of benzene and toluene are given in table 4.5. The identification of the active electrophile, and the elucidation of other aspects
of these reactions, will be of great interest.

TABLE 4.5   *Competitive nitrations*[45] *of toluene and benzene with
1-nitropyridinium tetrafluoroborates in acetonitrile at 25 °*

| Substituents in pyridinium salt | Relative rates | Isomer proportions (partial rate factors) | | | $\frac{1}{2}$ *o:p*-ratio |
		ortho	*meta*	*para*	
2-Me	36·5	63·8	3·2	33·0	0·965
		(69·9)	(3·5)	(72·3)	
2,6-Me$_2$	39·0	63·9	3·0	33·1	0·965
		(74·7)	(3·5)	(77·4)	
2,4,6-Me$_3$	41·4	63·1	3·1	33·8	0·93
		(78·5)	(3·8)	(84·0)	
4-MeO-2,6-Me$_2$	44·5	64·1	2·6	33·3	0·965
		(85·5)	(3·5)	(88·9)	
1,5-Dinitro-quinolinium*	13·2	62·1	2·1	35·8	0·865
		(24·6)	(0·8)	(28·4)	

* In nitromethane; the salt was not very soluble in acetonitrile.

REFERENCES

1. (a) Millen, D. J. (1950). *J. chem. Soc.* p. 2600.
    (b) Gillespie, R. J., Graham, J., Hughes, E. D., Ingold, C. K. & Peeling, E. R. A. (1950). *J. chem. Soc.* p. 2504.
2. Olah, G. A. & Kuhn, S. J. (1964). *Friedel–Crafts and Related Reactions* (ed. G. A. Olah), vol. 3, ch. 43. New York: Interscience.
3. Bachman, G. B. & Hokama, T. (1957). *J. Am. chem. Soc.* 79, 4370.
4. Evans, J. C., Rinn, H. W., Kuhn, S. J. & Olah, G. A. (1964). *Inorg. Chem.* 3, 857.
5. Fateley, W. G., Bent, H. A. & Crawford, B. (1959). *J. chem. Phys.* 31, 204.
6. Chédin, J. & Pradier, J. C. (1936). *C. r. hebd. Séanc. Acad. Sci., Paris* 203, 722.
7. Grison, E., Ericks, K. & de Vries, J. L. (1950). *Acta crystallogr.* 3, 290.
8. Susz, B. & Briner, E. (1935). *Helv. chem. Acta* 18, 378.
9. Fénéant, S. & Chédin, J. (1955). *Mém. Services chim. État* 40, 292.
10. Chédin, J. (1935). *C. r. hebd. Séanc. Acad. Sci., Paris* 201, 552.
11. Klemenc, A. & Schöller, A. (1924). *Z. anorg. allg. Chem.* 141, 231.

References

12. (a) Bamberger, E. (1894). *Ber. dt. chem. Ges.* **27**, 584.
    (b) Hoff, E. (1900). *Justus Liebigs Annln Chem.* **311**, 91.
    (c) Haines, L. B. & Adkins, H. (1925). *J. Am. chem. Soc.* **47**, 1419.
13. (a) Spreckels, E. (1919). *Ber. dt. chem. Ges.* **52**, 315.
    (b) Cooper, K. E. and Ingold, C. K. (1927). *J. chem. Soc.* p. 836.
14. Gold, V., Hughes, E. D., Ingold, C. K. & Williams, G. H. (1950). *J. chem. Soc.* p. 2452.
15. Bonner, T. G. & Hancock, R. A. (1966). *J. chem. Soc.* B, p. 972.
16. Bayliss, N. S., Dingle, R., Watts, D. W. & Wilkie, R. J. (1963). *Aust. J. Chem.* **16**, 933.
17. Singer, K. & Vamplew, P. A. (1956). *J. chem. Soc.* p. 3971.
18. Angus, W. R. & Leckie, A. H. (1935). *Proc. R. Soc.* A **150**, 615.
19. Deschamps, J. M. R. (1957). *C. r. hebd. Séanc. Acad. Sci., Paris* **245**, 1432.
20. Reed, R. I., quoted in Ref. 22.
21. Goulden, J. D. S. & Millen, D. J. (1950). *J. chem. Soc.* p. 2620. Millen, D. J. & Watson, D. (1957). *J. chem. Soc.* p. 1369.
22. Benford, G. A. & Ingold, C. K. (1938). *J. chem. Soc.* p. 929.
23. Hughes, E. D., Ingold, C. K. & Reed, R. I. (1950). *J. chem. Soc.* p. 2400.
24. Pinck, L. A. (1927). *J. Am. chem. Soc.* **49**, 2536.
25. (a) Arnall, F. (1923). *J. chem. Soc.* **123**, 3111; (1924). *J. chem. Soc.* **125**, 811.
    (b) Westheimer, F. H. & Kharasch, M. S. (1946). *J. Am. chem. Soc.* **68**, 1871.
    (c) Schramm, R. M. & Westheimer, F. H. (1948). *J. Am. chem. Soc.* **70**, 1782.
    (d) Lang, F. M. (1948). *C. r. hebd. Séanc. Acad. Sci., Paris* **226**, 1381; **227**, 849.
26. Bunton, C. A. with (a) Minkoff, G. J. & Reed, R. I. (1947). *J. chem. Soc.* 1416; (b) Hughes, E. D., Ingold, C. K., Jacobs, D. I. H., Jones, M. H., Minkoff, J. G. & Reed, R. I. (1950). *J. chem. Soc.* p. 2628; (c) Hughes, E. D., Minkoff, G. J. & Reed, R. I. (1946). *Nature, Lond.* **158**, 514.
27. Veibel, S. (1930). *Ber. dt. chem. Ges.* **63**, 1577.
28. Hughes, E. D., Ingold, C. K. & Ridd, J. H. (1958). *J. chem. Soc.* pp. 58, 65, 70, 77, 82, 88.
29. Blackall, E. L., Hughes, E. D. & Ingold, C. K. (1952). *J. chem. Soc.* p. 28.
30. Bonner, T. G. & Hancock, R. A. (1967). *Chem. Commun.* p. 780.
    Bonner, T. G., Hancock, R. A., Yousif, G. & (in part) Rolle, F. R. (1969). *J. chem. Soc.* B, p. 1237.
31. Hoggett, J. G., Moodie, R. B. & Schofield, K. (1969). *J. chem. Soc.* B, p. 1.
32. Schmeisser, M. & Elischer, S. (1952). *Z. Naturf.* 7 B, 583.
33. Olah, G. A. & Kuhn, S. J. (a) with Mlinkó, A. (1956). *J. chem. Soc.* p. 4257; (b) with Flood, S. H. (1961). *J. Am. chem. Soc.* **83**, 4571; (c) with Flood, S. H. (1961). *J. Am. chem. Soc.* **83**, 4581; (d) with Flood, S. H. & Evans, J. C. (1962). *J. Am. Chem. Soc.* **84**, 3687; (e) (1962). *J. Am. chem. Soc.* **84**, 3684.
34. Kuhn, S. J. & Olah, G. A. (1961). *J. Am. chem. Soc.* **83**, 4564.
35. Jones, J. & J. (1964). *Tetrahedron Lett.* p. 2117.
    Olah, G. A., Olah, J. A. & Overchuk, N. A. (1965). *J. org. Chem.* **30**, 3373.

36. Francis, A. W. (1926). *J. Am. chem. Soc.* **48**, 655.
37. (a) Christy, P. F., Ridd, J. H. & Stears, N. D. (1970). *J. chem. Soc.* B, p. 797.
    (b) Ridd, J. H. (1960). *Studies on Chemical Structure and Reactivity*, ed. J. H. Ridd, London: Methuen.
38. Ingold, C. K. & Shaw, F. R. (1927). *J. chem. Soc.* p. 2918.
39. Tolgyesi, W. S. (1965). *Can. J. Chem.* **43**, 343.
40. (a) Olah, G. A., Kuhn, S. J. & Carlson, C. G. (1963). *XIXth. Int. Congr. Pure and Applied Chem.* London, 1963. Abstr. of Papers A, pp. 1–81.
    (b) Olah, G. A. & Overchuk, N. A. (1965). *Can. J. Chem.* **43**, 3279.
41. Ritchie, C. P. & Win, H. (1964). *J. org. Chem.* **29**, 3093.
42. Ciaccio, L. L. & Marcus, R. A. (1962). *J. Am. chem. Soc.* **84**, 1838.
43. Olah, G. A., Kuhn, S. J. & Flood, S. H. (a) (1962). *J. Am. chem. Soc.* **84**, 1688; (b) with Hardie, B. A. (1964). *J. Am. chem. Soc.* **86**, 1039; (c) with Moffatt, M. E. & Overchuck, N. A. (1964). *J. Am. chem. Soc.* **86**, 1046.
44. Caille, S. Y. & Corriu, R. J. P. (1969). *Tetrahedron* **25**, 2005.
45. Cupas, C. A. & Pearson, R. L. (1968). *J. Am. chem. Soc.* **90**, 4742.
46. (a) Ingold, C. K., Lapworth, A., Rothstein, E. & Ward, D. (1931). *J. chem. Soc.* p. 1959.
    (b) Bird, M. L. & Ingold, C. K. (1938). *J. chem. Soc.* p. 918.
47. Knowles, J. R., Norman, R. O. C. & Radda, G. K. (1960). *J. chem. Soc.* p. 4885.
48. Roberts, J. D., Sanford, J. K., Sixma, F. L. J., Cerfontain, H. & Zagt, R. (1954). *J. Am. chem. Soc.* **76**, 4525.
49. Coombes, R. G., Moodie, R. B. & Schofield, K. (1968). *J. chem. Soc.* p. 800.
50. Brown, H. C. & Wirkkala, R. A. (1966). *J. Am. chem. Soc.* **88**, 1447.
51. Coombes, R. G., Crout, D. H. G., Hoggett, J. G., Moodie, R. B. & Schofield, K. (1970). *J. chem. Soc.* B, p. 347.
52. Hanna, S. B., Hunziker, E., Saito, T. & Zollinger, H. (1969). *Helv. chim. Acta* **52**, 1537.

5 Nitrating systems:
D. Benzoyl nitrate and systems formed from nitric acid and acetic anhydride

5.1 INTRODUCTION

Nitric acid and acetic anhydride react together to give, rapidly and almost quantitatively, acetyl nitrate; for this reason, nitrations effected by reagents prepared from nitric acid and acetic anhydride are considered together with nitrations effected with other acyl nitrates. We shall attempt to use the term 'acetyl nitrate' for solutions of acetyl nitrate in acetic anhydride which were prepared by adding nitric acid to acetic anhydride and allowing the two sufficient time to react together (about 30 min). It is not always possible to be so specific, for in much of the work that has been reported it has not been appreciated that the precise nature of the reagent used may depend upon its history.

It has long been known that, amongst organic solvents, acetic anhydride is particularly potent in nitration, and that reaction can be brought about under relatively mild conditions. For these reasons, and because aromatic compounds are easily soluble in mixtures of nitric acid and the solvent, these media have achieved considerable importance in quantitative studies of nitration.

The most notable studies are those of Ingold, on the orienting and activating properties of substituents in the benzene nucleus, and of Dewar on the reactivities of an extensive series of polynuclear aromatic and related compounds (§5.3.2). The former work was seminal in the foundation of the qualitative electronic theory of the relationship between structure and reactivity, and the latter is the most celebrated example of the more quantitative approaches to the same relationship (§7.2.3). Both of the series of investigations employed the competitive method, and were not concerned with the kinetics of reaction.

Nitration in acetic anhydride, or in solutions of dinitrogen pentoxide or of other acyl nitrates in carbon tetrachloride, has been associated with a higher ratio of o- to p-substitution in the reactions of certain compounds, than when the nitronium ion is the active species. This feature has been seen as indicating special mechanisms of nitration in some cases, and as providing information on the nature of the active electrophile (§5.3.4).

The kinetics of nitration in acetic anhydride are complicated. In addition to the initial reaction between nitric acid and the solvent, subsequent reactions occur which lead ultimately to the formation of tetranitromethane; furthermore, the observation that acetoxylation accompanies the nitration of the homologues of benzene adds to this complexity.

Another reason for discussing the mechanism of nitration in these media separately from that in inert organic solvents is that, as indicated above, the nature of the electrophile is not established, and has been the subject of controversy. The cases for the involvement of acetyl nitrate, protonated acetyl nitrate, dinitrogen pentoxide and the nitronium ion have been advocated.

These features serve to distinguish nitration in acetic anhydride from nitration in inert organic solvents. With other acyl nitrates less work has been done, and it is convenient to deal first with the case of benzoyl nitrate.

5.2 BENZOYL NITRATE

Nitration using this reagent was first investigated by Francis.[1] He showed that benzene and some of its homologues' bromobenzene, benzonitrile, benzoyl chloride, benzaldehyde and some related compounds, and phenol were mono-nitrated in solutions of benzoyl nitrate in carbon tetrachloride; anilines would not react cleanly and a series of naphthols yielded dinitro compounds. Further work on the orientation of substitution associated this reagent with higher proportions of *o*-substitution than that brought about by nitric acid; this point is discussed below (§5.3.4).

The kinetics of nitration of benzene in solutions at *c.* 20 °C in carbon tetrachloride have been investigated.[2] In the presence of an excess of benzene (*c.* 2–4 mol l⁻¹) the rate was kinetically of the first order in the concentration of benzoyl nitrate. The rate of reaction was depressed by the addition of benzoic anhydride, provided that some benzoic acid was present. This result suggested that benzoyl nitrate itself was not responsible for the nitration, but generated dinitrogen pentoxide according to the following equilibrium:

$$2BzONO_2 \rightleftharpoons N_2O_5 + Bz_2O.$$

The addition of pure benzoic anhydride did not affect the rate, therefore to explain the need for benzoic acid it was further suggested that the consumption of dinitrogen pentoxide was acid catalysed. Dinitrogen pentoxide, in concentrations similar to those of benzoyl nitrate used, nitrated benzene very rapidly even at $-20\ °C$, and furthermore this rate was not affected by the addition of pure benzoic anhydride. If, however, some benzoic acid were introduced with the anhydride, the rate was much reduced, and the nitrating power of the mixture resembled that of benzoyl nitrate. Using these mixtures it was shown that, in the presence of a small fixed concentration of benzoic acid, the observed rate varied approximately as the inverse power of the concentration of benzoic anhydride, and that in the presence of a fixed concentration of benzoic anhydride increasing concentrations of benzoic acid showed the same effect. The former observation indicates that the proportion of dinitrogen pentoxide formed in the first equilibrium is small. The facts are qualitatively interpreted in terms of the above equilibrium, with the effect of benzoic acid being expressed by the two following equilibria which provide a mechanism for the first:

$$N_2O_5 + BzOH \rightleftharpoons BzONO_2 + HNO_3,$$
$$Bz_2O + HNO_3 \rightleftharpoons BzONO_2 + BzOH.$$

Because of the chemical similarity between benzoyl nitrate and the acetyl nitrate which is formed in solutions of nitric acid in acetic anhydride, it is tempting to draw analogies between the mechanisms of nitration in such solutions and in solutions of benzoyl nitrate in carbon tetrachloride. Similarities do exist, such as the production by these reagents of higher proportions of *o*-substituted products from some substrates than are produced by nitronium ions, as already mentioned and further discussed below. Further, in solutions in carbon tetrachloride of acetyl nitrate or benzoyl nitrate, the addition of acetic anhydride and benzoic anhydride respectively reduces the rate of reaction, implying that dinitrogen pentoxide may also be involved in nitration in acetic anhydride.[2] However, for solutions in which acetic anhydride is also the solvent, the analogy should be drawn with caution, for in many ways the conditions are not comparable. Thus, carbon tetrachloride is a non-polar solvent, in which, as has been shown above, molecular dinitrogen pentoxide may be the effective electrophile, whereas acetic anhydride is a polar solvent which will more easily induce, or support, the heterolysis of this molecular species. Furthermore, the addition of benzoic acid anticatalyses nitration in solutions of benzoyl nitrate in carbon tetrachloride, but acetic acid has the opposite effect in nitration in acetic anhydride (§5.3.3).

5.3 SYSTEMS FORMED FROM NITRIC ACID AND ACETIC ANHYDRIDE

5.3.1 *The state of nitric acid in acetic anhydride*

Vandoni and Viala[3] examined the vapour pressures of mixtures of nitric acid in acetic anhydride, and concluded that from o to $\frac{1}{2}$ mole-fraction of nitric acid the solution consisted of acetyl nitrate, acetic acid and excess anhydride; in equimolar proportions the solution consisted of acetyl nitrate and acetic acid, and on increasing the fraction of nitric acid, dinitrogen pentoxide is formed, with a concentration which increases with the concomitant decrease in the concentration of acetyl nitrate.

A study of the Raman spectra of similar mixtures confirmed and extended these conclusions.[4] The existence of the following two equilibria was postulated:

$$Ac_2O + 2HNO_3 \rightleftharpoons N_2O_5 + 2AcOH,$$

$$N_2O_5 + Ac_2O \rightleftharpoons 2AcONO_2.$$

When acetic anhydride was in excess over nitric acid, acetyl nitrate and acetic acid were the only products. When the concentration of nitric acid was greater than 90 moles %, dinitrogen pentoxide, present as $(NO_2^+)(NO_3^-)$, was the major product and there were only small traces of acetyl nitrate. With lower concentrations of nitric acid the products were acetic acid, acetyl nitrate and dinitrogen pentoxide, the latter species being present as covalent molecules in this organic medium. A mixture of 2 moles of nitric acid and 1 mole of acetic anhydride has the same Raman spectrum as a solution of 1 mole of dinitrogen pentoxide in 2 moles of acetic acid.

An investigation of the infra-red spectra of mixtures of nitric acid and acetic anhydride supports these conclusions.[5] The concentration of nitronium ions, measured by the absorption band at 2380 cm^{-1}, was increased by the addition of small concentrations ($<$ 8 mol %) of acetic anhydride; further addition reduced the concentration. In low concentrations, acetic anhydride reacts to give ionized dinitrogen pentoxide, which on further addition of acetic anhydride is converted into the covalent molecule, and eventually into acetyl nitrate.

Evidence from the viscosities, densities, refractive indices[6] and measurements of the vapour pressure[7] of these mixtures also supports the above conclusions. Acetyl nitrate has been prepared from a mixture of acetic anhydride and dinitrogen pentoxide, and characterised, showing that the equilibria discussed do lead to the formation of that compound.[8]

The initial reaction between nitric acid and acetic anhydride is rapid at room temperature; nitric acid (0·05 mol l^{-1}) is reported to be converted into acetyl nitrate with a half-life of about 1 minute. This observation is consistent with the results of some preparative experiments, in which it was found that nitric acid could be precipitated quantitatively with urea from solutions of it in acetic anhydride at $-$10 °C, whereas similar solutions prepared at room temperature and cooled rapidly to $-$10 °C yielded only a part of their nitric acid (§5.3.2).[9a]

The following equilibrium has been investigated in detail:[10]

$$\text{Ac}_2\text{O} + \text{HNO}_3 \underset{k_2}{\overset{k_1}{\rightleftharpoons}} \text{AcONO}_2 + \text{AcOH}.$$

The equilibrium constant K, the rate constants k_1 and k_2 and the dependences of all these quantities on temperature were determined. In the absence of added acetic acid, the conversion of nitric acid into acetyl nitrate is almost quantitative. Therefore, to obtain at equilibrium a concentration of free nitric acid sufficiently high for accurate analysis, media were studied which contained appreciable concentrations (c. 4 mol l^{-1}) of acetic acid.

The mixture was prepared and allowed to achieve equilibrium; to it was added an excess of urea which caused the immediate precipitation as urea nitrate of the free nitric acid present. As a result of the sudden removal of the nitric acid from the mixture, the system underwent change to re-establish the equilibrium; however, the use of an excess of urea removed the nitric acid as it was produced from acetyl nitrate and acetic acid, and the consumption of acetyl nitrate proceeded to completion. Thus, by following the production of urea nitrate with the time from the addition of urea, the rate of the back reaction could be determined, and by extrapolating the results to zero time the equilibrium concentrations of free nitric acid and acetyl nitrate could be found. The rate of the forward reaction was not measured, but calculated from the expression $K = k_1 k_2^{-1}$. The values of the equilibrium and rate constants at several temperatures are listed in table 5.1.

TABLE 5.1   *Equilibrium and rate constants for the reaction*
$$Ac_2O + HNO_3 \rightleftharpoons AcONO_2 + AcOH$$

T/°C	K	$10^5 k_1/l\,mol^{-1}\,s^{-1}$	$10^5\,k_2/l\,mol^{-1}\,s^{-1}$
−10·0	.	3·81*	1·26*
0·0	2·46	16·9	6·86
5·0	2·00	21·4	10·70
10·0	1·67	.	.
10·3	.	27·9	16·3
15·0	1·40	32·4*	23·1*
20·0	1·18	41·2*	39·4*
25·0	1·00*	51·0*	51·0*

* These results were obtained by extrapolation.

More recent determinations[11b] of k_1 by the more direct method of observing changes in the absorbance of the solution at 290 nm gave values which were not in very good agreement with these earlier ones ($10^5\,k_1/l\,mol^{-1}\,s^{-1}$ at 4·0, 10·0 and 25·0 °C was 16·0, 30·0, and 95–120, respectively). The reaction was first order in the concentration of nitric acid ($[HNO_3] = 0\cdot04$–$0\cdot2$ mol l^{-1} at 25 ° C) and thus first-order overall.

In addition to the initial reaction between nitric acid and acetic anhydride, subsequent changes lead to the quantitative formation of tetranitromethane; in an equimolar mixture of nitric acid and acetic anhydride this reaction was half completed in 1–2 days.[12] An investigation[13] of the kinetics of this reaction showed it to have an induction period of 2–3 h for the solutions examined ([acetyl nitrate] = $0\cdot7$ mol l^{-1}), after which the rate adopted a form approximately of the first order with a half-life of about a day, close to that observed in the preparative experiment mentioned. In confirmation of this, recent workers have found the half-life of a solution at 25 °C of $0\cdot05$ mol l^{-1} of nitric acid to be about 2 days.[14b]

An observation which is relevant to the nitration of very reactive compounds in these media (§5.3.3) is that mixtures of nitric acid and acetic anhydride develop nitrous acid on standing. In a solution ($[HNO_3] = 0\cdot7$ mol l^{-1}) at 25 °C the concentration of nitrous acid is reported to increase from 3×10^{-4} mol l^{-1} to 5×10^{-2} mol l^{-1} in a few hours.[13] These solutions did not contain any aromatic compound, and therefore the observation is not related to the occurrence of acetoxylation which, it is maintained, produces equivalent amounts of nitrous acid.[14a]

5.3.2 *Characteristics of the system as nitrating reagents*

Wibaut, who introduced the competitive method for determining reactivities (his experiments with toluene, benzene and chlorobenzene were performed under heterogeneous conditions and were not successful), pointed out that solutions of nitric acid in acetic anhydride are useful in making comparisons of reactivities because aromatic compounds are soluble in them.[15]

Ingold and his co-workers used the competitive method in their experiments, in which nitration was brought about in acetic anhydride. Typically, the reaction solutions in these experiments contained 0·8–1·4 mol l^{-1} of nitric acid, and the reaction time, depending on the reactivities of the compounds and the temperature, was 0·5–10 h. Results were obtained for the reactivities of toluene,[16a,b] ethyl benzoate,[16c] the halogenobenzenes,[16d] ethyl phenyl acetate and benzyl chloride.[16e] Some of these and some later results are summarized in table 5.2. Results for the halogenobenzenes and nitrobiphenyls are discussed later (§9.1.4, 10.1), and those for a series of benzylic compounds in §5.3.4.

Dewar and his co-workers, as mentioned above, investigated the reactivities of a number of polycyclic aromatic compounds because such compounds could provide data especially suitable for comparison with theoretical predictions (§7.2.3). This work was extended to include some compounds related to biphenyl. The results were obtained by successively compounding pairs of results from competitive nitrations to obtain a scale of reactivities relative to that of benzene. Because the compounds studied were very reactive, the concentrations of nitric acid used were relatively small, being 0·18 mol l^{-1} in the comparison of benzene with naphthalene, 5×10^{-3} mol l^{-1} when naphthalene and anthanthrene were compared, and 3×10^{-3} mol l^{-1} in the experiments with diphenylamine and carbazole. The observed partial rate factors are collected in table 5.3. Use of the competitive method in these experiments makes them of little value as sources of information about the mechanisms of the substitutions which occurred; this shortcoming is important because in the experiments fuming nitric acid was used, rather than nitric acid free of nitrous acid, and with the most reactive compounds this leads to a

TABLE 5.2 *Nitrations of mono-substituted derivatives of benzene*
*in solutions of acetyl nitrate in acetic anhydride**

Compound	Temp °C	Partial rate factors			Ref.
		f_o	f_m	f_p	
Benzyl chloride	25	0·29	0·14	0·95	16e
tert-Butylbenzene†	0	4·5	3·0	75·5	17
Cinnamic acid	25	(Rate relative to that for benzene: 0·111)			18
Ethylbenzene	25	31	2·3	70	19
Ethyl benzoate	18	$2·6 \times 10^{-3}$	$7·9 \times 10^{-3}$	9×10^{-4}	16c
Ethyl phenyl acetate	25	4·6	1·2	10·4	16e
iso-Propylbenzene	25	15	2·4	72	19
Toluene	0	47	3	62	16b
	0	50	1·3	60	19
	25	46·5	2·1	48·5	17
	25	42	1·9	62·6	20
Trimethylsilylbenzene	0–10	1·3	2·0	3·1	21

* See also Table 4.2. † For nitration with $MeNO_2$–HNO_3 at 25 °C and with HNO_3–90 % aq. AcOH at 45°, f_o, f_m and f_p were 5·5, 3·7 and 72, and 5·5, 4·0, and 75, respectively.[17, 22]

change of mechanism (§5.3.3). As a consequence, the very high partial rate factors reported for some compounds (table 5.3) are not appropriate for comparison with those for the less reactive compounds. An upper limit to the significance of the data for these latter compounds is set by the observation of a limiting rate of nitration upon encounter for the primary mechanism of nitration (§5.3.3). The observations are important in connection with theoretical treatments of reactivity (§7.2.3).

Certain features of the addition of acetyl nitrate to olefins in acetic anhydride may be relevant to the mechanism of aromatic nitration by this reagent. The rapid reaction results in predominantly *cis*-addition to yield a mixture of the β-nitro-acetate and β-nitro-nitrate.[9] The reaction was facilitated by the addition of sulphuric acid, in which case the yield of β-nitro-nitrate was reduced, whereas the addition of sodium nitrate favoured the formation of this compound over that of the acetate. As already mentioned (§5.3.1), a solution of nitric acid (c. 1·6 mol l⁻¹) in acetic anhydride prepared at − 10 °C would yield 95–97 % of the nitric acid by precipitation with urea, whereas from a similar solution prepared at 20–25 °C and cooled rapidly to − 10 °C only 30 % of the acid could be recovered. The difference between these values was attributed to the formation of acetyl nitrate. A solution prepared at room

TABLE 5.3   *The nitration of polycyclic aromatic compounds in solutions of acetyl nitrate in acetic anhydride*

Compound	Position of substitution	Partial rate factor*	Ref.
Anthanthrene	6 (?)	156 000 (185 600)	23 b,d
Benzo[a]pyrene	6	108 000 (126 000)	23 b,d
Biphenyl	2	30 (18·7)	23 b,d
	4	18 (11·1)	
Chrysene	2 (6)	3 500	23 b,d
Coronene	1	1 150	23 d
Naphthalene	1	470	23 a,d
	2	50	
Perylene	3	77 000	23 a,d
Phenanthrene	1	360	23 a–c
	2	92	
	3	300	
	4	79	
	9	490	
Pyrene	1	17 000	23 d
Triphenylene	1	600	23 d
	2	600	
Carbazole	1	32 100 (31 080)	24 b
	2	1 100 (2 220)	
	3	77 600 (77 700)	
Dibenzofuran	1	47	24 a,b
	2	94	
	3	94	
Diphenylamine	2	831 000	24 b
	4	575 000	
Diphenyl ether	2	117	24 b
	4	231	
Diphenylmethane	2	13	24 b
	4	32	
Fluoranthene	1	330	25
	3	1 365	
	7	564	
	8	846	
Fluorene	2	2 040 (209·8)	24 b
	3	60 (6·1)	
	4	940 (88·2)	
10-Methyl-10,9-borazarophenanthrene	6	937 000	23 e
	8	2 060 000	

* For an explanation of the figures in parentheses see §10.3.

temperature and cooled to − 15 °C was found to be considerably more powerful, in the reaction with alkenes, than a corresponding solution prepared and maintained at − 10 °C.

Protonated acetyl nitrate and, to a smaller extent, acetyl nitrate were thought to be responsible for the formation of the β-nitro-acetate, whereas the β-nitro-nitrate was thought to be formed from the small concentration of dinitrogen pentoxide present in these solutions. Sulphuric acid, by converting acetyl nitrate into its more reactive protonated form, enhanced the formation of the nitro-acetate, and added salts brought about the opposite effect by abstracting a proton from the protonated species, supposed to exist in small concentrations in the absence of strong acid.

It was shown that in preparative experiments sulphuric acid markedly catalysed, and acetate ions markedly anticatalysed the nitration of anisole.*

The authors of this work were concerned chiefly with additions to alkenes, and evidence about the mechanism of aromatic nitration arises by analogy. Certain aspects of their work have been repeated to investigate whether the nitration of aromatic compounds shows the same phenomena[26] (§5.3.6). It was shown that solutions of acetyl nitrate in acetic anhydride were more powerful nitrating media for anisole and biphenyl than the corresponding solutions of nitric acid in which acetyl nitrate had not been formed; furthermore, it appeared that the formation of acetyl nitrate was faster when 95–98% nitric acid was used than when 70% nitric acid was used.

5.3.3 *Kinetic studies*

First-order nitrations. The kinetics of nitrations in solutions of acetyl nitrate in acetic anhydride were first investigated by Wibaut.[15] He obtained evidence for a second-order rate law, but this was subsequently disproved.[27] A more detailed study was made using benzene, toluene, chloro- and bromo-benzene.[13] The rate of nitration of benzene was found to be of the first order in the concentration of aromatic and third order in the concentration of acetyl nitrate; the latter conclusion disagrees with later work (see below). Nitration in solutions containing similar concentrations of acetyl nitrate in acetic acid was too slow to measure, but was accelerated slightly by the addition of more acetic anhydride. Similar solutions in carbon tetrachloride nitrated benzene too quickly, and the concentration of acetyl nitrate had to be reduced from 0·7 to 0·1 mol l^{-1} to permit the observation of a rate similar to that which the more concentrated solution yields in acetic anhydride.

---

* The effect of acetate ions cannot be distinguished from that of nitrate ions, which would be produced when acetate was added to the medium.

The addition of acetic anhydride to these solutions retarded the rate, but it has subsequently been shown that in the absence of any acetic anhydride the rate was very low.[2]

The rates of nitration of benzene in solutions at 25 °C containing 0·4–2·0 mol l⁻¹ of acetyl nitrate in acetic anhydride have been determined.[28] The rates accord with the following kinetic law:

$$\text{rate} = k \text{ [benzene] [acetyl nitrate].}[2]$$

The nitric acid used in this work contained 10 % of water, which introduced a considerable proportion of acetic acid into the medium. Further dilution of the solvent with acetic acid up to a concentration of 50 moles % had no effect on the rate, but the addition of yet more acetic acid decreased the rate, and in the absence of acetic anhydride there was no observed reaction. It was supposed from these results that the adventitious acetic acid would have no effect. The rate coefficients of the nitration diminished rapidly with time: in one experiment the value of k was reduced by a factor of 2 in 1 h. Corrected values were obtained by extrapolation to zero time. The author ascribed the decrease to the conversion of acetyl nitrate into tetranitromethane, but this conversion cannot be the explanation because independent studies agree in concluding that it is too slow (§5.3.1).

Recent experiments[11] have shown that the concentration of aromatic compound needed to maintain zeroth-order kinetics (see below) was much greater than for nitrations with solutions of nitric acid in some inert organic solvents; reactions which were first order in the concentration of the aromatic were obtained when [ArH] $< c.$ 2×10^{-2} mol l⁻¹.

Under these first-order conditions the rates of nitration of a number of compounds with acetyl nitrate in acetic anhydride have been determined. The data show that the rates of nitration of compounds bearing activating substituents reach a limit;[11a] by analogy with the similar phenomenon shown in nitration in aqueous sulphuric and perchloric acids (§2.5) and in solutions of nitric acid in sulpholan and nitromethane (§3.3), this limit has been taken to be the rate of encounter of the nitrating entity with the aromatic molecule.

Zeroth-order nitrations. The rates of nitration at 25 °C in solutions of acetyl nitrate (6×10^{-3} – 0·22 mol l⁻¹) in acetic anhydride of *o*- and *m*-xylene,[11, 14b] and anisole and mesitylene[11] were independent of the concentration and nature of the aromatic compound provided that

[ArH] $\nleqslant$ *c.* 10^{-1} mol l^{-1}. The reactions of *o*- and *m*-xylene were complicated by acetoxylation (see below).[11, 14]

The dependence of the zeroth-order rate constants on the concentration of acetyl nitrate is shown in fig. 5.1; in the absence of added acetic
acid the rate increases according to the third power of the concentration
of acetyl nitrate, but when acetic acid is added the dependence becomes

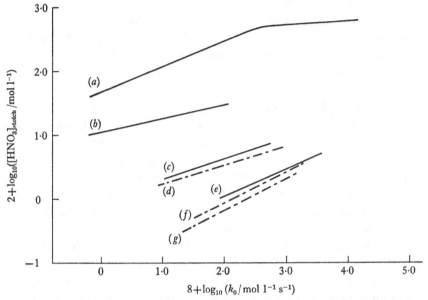

Fig. 5.1. Zeroth-order rates of nitration with acetyl nitrate compared with those for
other systems. (*a*) HNO$_3$/ sulpholan, (*b*) HNO$_3$/CCl$_4$, (*c*) AcONO$_2$/Ac$_2$O/O % AcOH,[14b]
(*d*) AcONO$_2$/Ac$_2$O/O % AcOH/[mesitylene] = 0·8 mol l^{-1},[11] (*e*) AcONO$_2$/Ac$_2$O/
[AcOH] = 2·2 mol l^{-1},[14b] (*f*) AcONO$_2$/Ac$_2$O/[AcOH] = 1·96 mol l^{-1} and (*g*) AcONO$_2$/
Ac$_2$O/[AcOH] 3·91 mol l^{-1}/[mesitylene] = 0·8 mol l^{-1}.[11]

of the second order.[11b, d, 14a] In fig. 5.1 the results are also compared with
those for nitration in solutions of nitric acid in sulpholan and in carbon
tetrachloride (§3.2).

The results in fig. 5.1 show that zeroth-order rates of nitration in
solutions of acetyl nitrate in acetic anhydride are much greater than the
corresponding rates in solutions in inert organic solvents of nitric acid
of the same stoichiometric concentration as that of acetyl nitrate. Thus,
for corresponding concentrations of nitric acid and acetyl nitrate,
nitration in acetic anhydride is *c.* 5×10^5 and 10^4 times faster than nitration in sulpholan and nitromethane respectively. This fact, and the fact
that the fraction of free nitric acid in solutions of acetyl nitrate in acetic

anhydride is small[11] (see § 5.3.1) makes it improbable that the mechanism for the formation of the active species in these solutions is the same as that operating in solutions of nitric acid in inert organic solvents. Therefore it seems justifiable to conclude that the mechanism of nitration in solutions of acetyl nitrate involves acetyl nitrate at some stage, or some other species not present in significant concentrations in solutions of nitric acid in inert organic solvents. The above comparisons were made with sulpholan and nitromethane rather than with the non-polar solvent carbon tetrachloride, because it has been shown (§3.2.1) that in that solvent molecules of nitric acid congregate, and that such behaviour gives to nitration in these solutions characteristics different from those observed in more polar solvents; acetic anhydride is a polar solvent in which such a phenomenon would not be expected to occur.

Nitrations of the zeroth order are maintained with much greater difficulty in solutions of acetyl nitrate in acetic anhydride than in solutions of nitric acid in inert organic solvents, as has already been mentioned. Thus, in the former solutions, the rates of nitration of mesitylene deviated towards a dependence on the first power of its concentration when this was $< c.$ 0·05–0·1 mol l^{-1}, whereas in nitration with nitric acid in sulpholan, zeroth-order kinetics could be observed in solutions containing as little as 10^{-4} mol l^{-1} of mesitylene (§3.2.1).

The observation of nitration at a rate independent of the concentration and the nature of the aromatic means only that the effective nitrating species is formed slowly in a step which does not involve the aromatic. The fact that the rates of zeroth-order nitration under comparable conditions in solutions of nitric acid in acetic acid, sulpholan and nitromethane differed by at most a factor of 50 indicated that the slow step in these three cases was the same, and that the solvents had no chemical involvement in this step. The dissimilarity in the rate between these three cases and nitration with acetyl nitrate in acetic anhydride argues against a common mechanism, and indeed it is not required from evidence about zeroth-order rates alone that in the latter solutions the slow step should involve the formation of the nitronium ion.

Acetoxylation and nitration. It has already been mentioned that *o*- and *m*-xylene are acetoxylated as well as nitrated by solutions of acetyl nitrate in acetic anhydride. This occurs with some other homologues of benzene,[14a] and with methyl phenethyl ether,[11b] but not with anisole, mesitylene or naphthalene.[11b] Results are given in table 5.4.

o- and *m*-xylene were selected for detailed kinetic investigation.[14b] The runs were performed at 25 °C using a concentration of acetyl nitrate in the range 0·1–0·01 mol l⁻¹, with an excess of aromatic compound (*c.* 0·1–0·5 mol l⁻¹). The nitration and acetoxylation reactions were followed simultaneously, and in all the experiments undertaken these two reactions maintained the same relative importance.

The effects of added species. The rate of nitration of benzene, according to a rate law kinetically of the first order in the concentration of aromatic, was reduced by sodium nitrate, a concentration of 10^{-3} mol l⁻¹ of the latter retarding nitration by a factor of about 4.[11c, 28] Lithium nitrate anticatalysed the nitration and acetoxylation of *o*-xylene in solutions of acetyl nitrate in acetic anhydride. The presence of 6×10^{-4} mol l⁻¹ of nitrate reduced the rate by a factor of 4, and modified the kinetic form of the nitration from a zeroth-order dependence on the concentration of aromatic towards a first-order dependence. However, the ratio of acetoxylation to nitration remained constant.[14b] Small concentrations of sodium nitrate similarly depressed the rate of nitration of anisole and again modified the reaction away from zeroth to first-order dependence on the concentration of the aromatic.[11b]

The addition of sulphuric acid increased the rate of nitration of benzene, and under the influence of this additive the rate became proportional to the first powers of the concentrations of aromatic, acetyl nitrate and sulphuric acid. Sulphuric acid markedly catalysed the zeroth-order nitration and acetoxylation of *o*-xylene without affecting the kinetic form of the reaction.[14b]

In the nitration and acetoxylation of *o*-xylene the addition of acetic acid increased the rate in proportion to its concentration, the presence of 3·0 mol l⁻¹ accelerating the rate by a factor of 30. In the presence of a substantial concentration (2·2 mol l⁻¹) of acetic acid the rate of reaction obeyed the following kinetic expression:[14b]

$$\text{rate} = k \, [\text{acetyl nitrate}]^2 \, [\text{acetic acid}].$$

Similarly, acetic acid catalysed the zeroth-order nitration of mesitylene without affecting the kinetic form [11b, d]

The nitration of very reactive compounds. Under the conditions where less-reactive compounds were nitrated according to a first-order law the nitrations of anthanthrene, diphenylamine, phenol, and resorcinol were

TABLE 5.4 *Yields (moles %) in the nitration and acetoxylation of some derivatives of benzene at 25 °C*

Compound	Nitro-compounds				Acetoxy-compounds			
	2	3	4	5	2	3	4	5
Hemimellitene[14c]			46.4±2.3	7.7±0.3			10.6±1.1	35.3±3.3
Methyl phenethyl ether[11b]*	60	5	33				2†	
	59	5	34				2†	
Pseudocumene[14c]		10.0±2.0		49.5±3.2		10.3±2.2	3.1±0.7	25.6±1.3
Toluene[14c]	58.2±2.4	2.8±1	35.9±1.8					
Toluene[11b]	59	3	35				3	
o-Xylene[14c]		13.5±0.7	28.0±1.3				58.6±1.8	
o-Xylene[11b]		22	37				41	
o-Xylene[11b]‡		24	36				40	
o-Xylene[11b]§		24	36				40	
m-Xylene[14c]	14.7±1.6		84.7±1.6				0.7±0.8	
m-Xylene[11b]	18		81				0.9||	

* The second row of results refers to an experiment using a freshly prepared solution of nitric acid (1.0 mol l^{-1}) in acetic anhydride.
† Not identified.
‡ At 4.0 °C.
§ At 4.0 °C using a freshly prepared solution of nitric acid (0.14 mol l^{-1}) in acetic anhydride for 20 min; under these conditions the half-life for the conversion of nitric acid into acetyl nitrate was 7 min.
|| 0.4 mole % of an unidentified product were also formed.[11b]

not always kinetically of the first order in the concentration of the aromatic, and those of phenol and resorcinol were prone to auto-catalysis.[11a] These observations relate to solutions prepared from pure nitric acid in which $[HNO_2] < c.\ 10^{-4}$ mol l^{-1}. If the nitrating solutions were kept for several hours before use [when nitrous acid is developed (§5.3.1)], or if 'fuming' nitric acid were used in their preparation, the rates of nitration of these four compounds, and of most of the compounds which react at the limiting rate when pure nitric acid was used, were accelerated enormously. It is evident that small concentrations of nitrous acid have a very strong effect.

The evidence outlined strongly suggests that nitration *via* nitrosation accompanies the general mechanism of nitration in these media in the reactions of very reactive compounds.[11a] Proof that phenol, even in solutions prepared from pure nitric acid, underwent nitration by a special mechanism came from examining rates of reaction of phenol and mesitylene under zeroth-order conditions. The variation in the initial rates with the concentration of aromatic (fig. 5.2) shows that mesitylene (0·2–0·4 mol l^{-1}) reacts at the zeroth-order rate, whereas phenol is nitrated considerably faster by a process which is first order in the concentration of aromatic. It is noteworthy that in these solutions the concentration of nitrous acid was below the level of detection ($< c.\ 5 \times 10^{-5}$–10^{-4} mol l^{-1}).[11b]

The orientation of substitution into phenol was strongly dependent on the conditions of reaction (see below).

Despite the fact that solutions of acetyl nitrate prepared from purified nitric acid contained no detectable nitrous acid, the sensitivity of the rates of nitration of very reactive compounds to nitrous acid demonstrated in this work is so great that concentrations of nitrous acid below the detectable level could produce considerable catalytic effects. However, because the concentration of nitrous acid in these solutions is unknown the possibility cannot absolutely be excluded that the special mechanism is nitration by a relatively unreactive electrophile. Whatever the nature of the supervenient reaction, it is clear that there is at least a dichotomy in the mechanism of nitration for very reactive compounds, and that, unless the contributions of the separate mechanisms can be distinguished, quantitative comparisons of reactivity are meaningless.

This qualification must be applied to the results of Dewar and his co-workers relating to the reaction of a series of polynuclear aromatic compounds with solutions of nitric acid in acetic anhydride at o °C

(§5.3.2). At this temperature, and provided that the concentration of acetic acid in the acetic anhydride was small, the conversion of nitric acid into acetyl nitrate would have had a half-life of 7–10 min. The description of the experimental method[23] makes it clear that the solutions used by Dewar in this work contained acetyl nitrate over the vast majority of the reaction. Therefore it must be supposed that in this

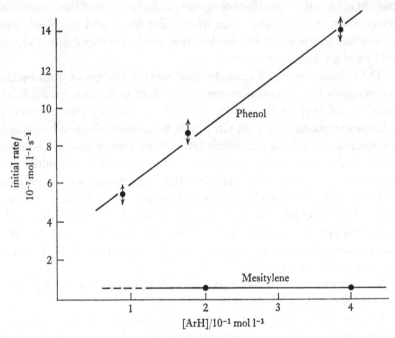

Fig. 5.2. Initial rates of nitration of mesitylene and phenol under zeroth-order conditions.[11b] Temperature 25 °C. [AcONO$_2$] = *c*. 7 × 10^{-3} mol l^{-1}. [HNO$_2$] < 10^{-4} mol l^{-1}.

work the additional mechanism of nitration became superimposed on a more general mechanism in such a way as to create a misleading impression of continuity in the reactivities of the compounds investigated. That the nitration solutions used in this work were prepared from 'fuming' nitric acid, evidently without purification, strengthens this conclusion, for in 'fuming' nitric acid the concentration of nitrous acid is relatively high (*c*. 0·06 mol l^{-1}).

5.3.4 *o:p-Ratios*

Nitrations in acetic anhydride, or in solutions containing benzoyl nitrate (§5.2) or dinitrogen pentoxide (§4.2.3) have long been associated with the formation from some aromatics of higher proportions of *o*-nitro-compounds than are formed under other conditions.

Several explanations have been suggested for these results, such as a general change in the nitrating species. This explanation has been regarded as unlikely,[29] for the change in orientation of substitution is not always very significant as, for example, in the cases of such compounds as toluene,[30] *tert*-butylbenzene and the halogenobenzenes (§5.3.2) [cf. the case of chlorobenzene reacting with dinitrogen pentoxide (§4.2.3)].

The change has been attributed to an electrostatic effect arising from the dipole of the substituent.[31] In the cases of anisole and acetanilide the positive end of the dipole is directed away from the ring, resulting in the *o*-position's being negatively polarized with respect to the *p*-position, and therefore more susceptible to electrophilic attack. This electrostatic effect was considered to be stronger in acetic anhydride, a solvent of lower dielectric constant, than in mixed acid. However, the results (see below) show that this effect does not occur in acetic acid, which has an even smaller dielectric constant than has acetic anhydride. For the halogenobenzenes, the dipole of the carbon–halogen bond is oriented in the opposite direction, and therefore the same change in solvent should bring about the opposite effect. Chlorobenzene and bromobenzene were found[31] to yield in acetic anhydride 10% and 90%, and 25% and 75% of *o*- and *p*-nitro-compounds respectively, and in concentrated nitric acid the proportions were 30% and 70%, and 38% and 62% respectively. These results are in accord with the electrostatic interpretation, but at variance with results obtained by other workers (§9.1.4).

This hypothesis has, however, been supported.[32] The *o:p*-ratio in chlorobenzene was found to be lower when acetic anhydride was the solvent, than when nitric acid or mixed acids were used. The ratio was still further reduced by the introduction into the solution of an even less polar solvent such as carbon tetrachloride, and was increased by the addition of a polar solvent such as acetonitrile. The orientation of substitution in toluene in which the substituent does not posses a strong dipole was found to be independent of the conditions used. The author concluded that the changes in orientation which accompany the change in the solvent arise largely from differences in polarities, although in the cases of anisole and acetonitrile other factors such as the electronic and steric effects associated with changes in hydrogen-bonding are likely to be important. The point was also made[32] that if protonated acetyl nitrate were considered to be a tightly solvated nitronium ion, as illustrated, then those features of nitration in these solutions which seem

$$CH_3 - C \overset{\displaystyle O \cdots H}{\underset{\displaystyle O : \overset{+}{N} \diagdown O}{\diagup}} \quad O$$

to require the operation of protonated acetyl nitrate could be reconciled with the fact that, for many compounds, relative rates and the orientation of substitution are not very different from those brought about by the nitronium ion.

Biphenyl is a compound which raises problems as regards the orientation of nitration produced by different reagents. This compound is discussed later (§ 10.1).

We shall now examine critically some compounds which have been particularly widely discussed in the present context.

Anisole and acetanilide. Some results for anisole are given in table 5.5. Halvarson and Melander[30] showed that o-nitration of anisole with benzoyl nitrate (benzoyl chloride and silver nitrate) in acetonitrile was not associated with a kinetic isotope effect (§6.4). They offered two explanations of the high o:p-ratios observed with nitrations using benzoyl nitrate or acetyl nitrate (table 5.5, expts. 8–9) as compared with those using reagents normally operating through the nitronium ion (table 5.5, expts. 1–7); either the initial attack of the nitrating species took place at oxygen, followed by an intramolecular rearrangement to the o-nitro-compound, or the orientation observed using acetyl nitrate was the normal one, and the results of experiments carried out using more strongly acidic media had been affected by nitrosation.

Later experiments do not allow a clear choice between these alternatives. The high proportion of o-isomer formed when nitration is effected with acetyl nitrate in acetic anhydride is confirmed by the results of expts. 10–14 (table 5.5). The use of fuming, rather than pure nitric acid, in the preparation of the reagent, which may lead to nitration

TABLE 5.5   *Proportions of isomers formed in the nitration*
of anisole

Reagent	Temp. °C	Isomers (%)			Ref.
		ortho	*meta*	*para*	
1 HNO$_3$, H$_2$SO$_4$ (1:1)	45	31	2	67	33
2 HNO$_3$ (d 1·42)	45	40	1	58	33
3 HNO$_3$ (25%), AcOH	65	44	2	54	33
4 HNO$_3$ (d 1·42)	45	43	0*	57	34
5 HNO$_3$, H$_2$SO$_4$ (1:1)	45	40	0*	60	34
6 HNO$_3$ (3 mol l^{-1}), AcOH†	20	30·5	0*	69·5	34
7 HNO$_3$ (6 mol l^{-1}), AcOH‡	20	34	0*	66	34
8 AcONO$_2$, Ac$_2$O	10	71	<0·5	28	30
9 BzONO$_2$, MeCN	0	75	<0·5	25	30
10 AcONO$_2$ (7·3 × 10^{-2} mol l^{-1}), Ac$_2$O§	25	74	.	26	11
11 AcONO$_2$ (6·3 × 10^{-2} mol l^{-1}), Ac$_2$O§	25	72	.	28	11
		70·5	.	29·5	11
12 AcONO$_2$ (6·8 × 10^{-2} mol l^{-1}), Ac$_2$O, [NO$_3^-$] = 10^{-3} mol l^{-1}§	25	71	.	29	11
13 AcONO$_2$ (3·4 × 10^{-2} mol l^{-1}), Ac$_2$O, [H$_2$SO$_4$] = 10^{-4} mol l^{-1}§	25	70	.	30	11
14 AcONO$_2$(0·25 mol l^{-1}), Ac$_2$O‖	25	71	.	29	11
15 AcONO$_2$ (0·25 mol l^{-1}), Ac$_2$O, 'fuming' nitric acid used‖	25	68	.	32	11
16 HNO$_3$ (6·55 mol l^{-1}), sulpholan [urea] = 4 × 10^{-2} mol l^{-1} ¶	25	68·5	.	31·5	11
17 HNO$_3$ (3 × 10^{-3} mol l^{-1}), 68·9% H$_2$SO$_4$. [urea] = 2 × 10^{-2} mol l^{-1}	25	60	.	40	11
18 HNO$_3$ (3 × 10^{-2} mol l^{-1}), 65·9% H$_2$SO$_4$. [urea] = 2 × 10^{-2} mol l^{-1}**	25	61	.	39	11
19 HNO$_3$ (3 × 10^{-2} mol l^{-1}), 65·9% H$_2$SO$_4$. [NaNO$_2$] = c. 4 × 10^{-2} mol l^{-1}	25	6	.	94	11

* A careful search was made for this isomer.
† [HNO$_2$] = 3 × 10^{-2} mol l^{-1}.
‡ [HNO$_2$] = 0·12 mol l^{-1}.
§ [Anisole] = 0·4 mol l^{-1}; zeroth-order reactions, except for that with added nitrate, which was of the first order with respect to the concentration of anisole.
‖ [Anisole] = 2 × 10^{-2} mol l^{-1}; first-order reactions. For the experiment using pure nitric acid the half-life was about 1 min, but for that using fuming nitric acid reaction was complete in < 30 s.
¶ [Anisole] = 4 × 10^{-2} mol l^{-1}; zeroth-order reaction (k_0 = 7·9 × 10^{-6} mol l^{-1} s^{-1}) with the same rate constant as that for mesitylene (7·3 × 10^{-6} mol l^{-1} s^{-1}).
** k_2 = 1·2 × 10^{-1} l mol^{-1} s^{-1}. Under the same conditions k_2 = c. 2 × 10^{-1} l mol^{-1} s^{-1} for mesitylene.[35]

through nitrosation, causes little change in the isomer ratio. Nitration of anisole with nitric acid in sulpholan (table 5.5, expt. 16), a nitronium ion reaction, gave a very similar result, but reaction in aqueous sulphuric acid, also a nitronium ion reaction, produces a small change towards p-substitution. Hydrogen bonding of the anisole in the aqueous acidic medium might be influential in this case. That nitration through nitrosation can have a marked effect on the isomer ratios is clearly shown by experiment 19 (table 5.5); the large swing towards p-substitution brought about by addition of sodium nitrite in this case leads one to speculate that the earlier results (table 5.5, expts. 1–9) may have been influenced by nitrosation.

Acetanilide is another compound which has attracted attention because of the way in which the orientation of its nitration depends upon the reagent. Results are given in table 5.6. They have usually been discussed in terms of some sort of special interaction between the nitrating agent (assumed to be of the type NO_2X) operating in solutions of acetyl nitrate in acetic anhydride and the oxygen or nitrogen atom of the acetylamino-group, which is supposed to lead to a high proportion of o-substitution.[29] One form of this sort of explanation is derived from work on methyl phenethyl ether (see below); it supposes that with acetyl nitrate in acetic anhydride there is, superimposed on nitronium ion nitration, a reaction with dinitrogen pentoxide which leads specifically to o-substitution.[39] Nitronium tetrafluoroborate is regarded as being a reagent which does not operate through free nitronium ions and thus, like the effective electrophile from acetyl nitrate, as producing a high $o:p$-ratio by one of these special mechanisms.[37]

Without further studies little weight can be given to these ideas. In particular there is the possibility that with acetanilide, as with anisole, nitrosation is of some importance, and further with nitrations in sulphuric acid the effect of protonation of the substrate needs quantitative evaluation. The possibility that the latter factor may be important has been recognised,[37] and it may account for the difference between nitration in sulphuric acid and nitration with nitronium tetrafluoroborate.

Phenol. The change in the orientation of substitution into phenol as a result of the superimposition of nitrosation on nitration is a well-established phenomenon.[40] In aqueous sulphuric acid it leads to a change from the production of 73 % of o-nitrophenol under nitrating

TABLE 5.6   *Proportions of isomers formed in the nitration of acetanilide**

Reagent	Temp. °C	Isomers %			o:p Ratio	Ref.
		ortho	*meta*	*para*		
HNO$_3$, H$_2$SO$_4$ (1:1)	20	19·4	2·1†	78·5	0·25	36
HNO$_3$, H$_2$SO$_4$ (1:2·3)	0	.	.	75	0·05	37
90% HNO$_3$	−20	23·5	0	76·5	0·30	38
80% HNO$_3$	−20	40·7	0	59·3	0·69	38
AcONO$_2$, Ac$_2$O	20	67·8	2·5†	29·7	2·3	36
AcONO$_2$, Ac$_2$O	0	72	.	.	4·2	37
NO$_2$BF$_4$, MeCN	−30 to −10	.	.	.	3·5 ± 0·5	37

* Data for substituted acetanilides are also available.[37]
† These values are probably too high.[34]

conditions to the production of 9% under predominantly nitrosating conditions (table 5.7, expts. 1–3). The apparently contradictory circumstances shown in expts. 4–10 (table 5.7) almost certainly arise from a change in nitrosating agent; it has been suggested that in expts. 4–6 the nitrosonium ion is dominant, and in expts. 7–10 dinitrogen tetroxide itself is the nitrosating agent. Overall the results are made difficult to interpret by ignorance of the proportions of substitution resulting from nitration and from nitrosation.

In solutions of acetyl nitrate in acetic anhydride, prepared from purified nitric acid, the *o*:*p*-ratio increases slightly with increasing concentrations of acetyl nitrate (table 5.7, expts. 11, 13, 16). The use of fuming nitric acid in the preparation of the acetyl nitrate considerably accelerates the rates of reaction and also increases the proportion of *o*-substitution (table 5.7, expts. 12, 15, 18). These effects resemble, but are much stronger than the corresponding effects in nitrations with solutions of nitric acid in acetic acid containing dinitrogen tetroxide.

The change in the distribution of substitution accompanying marked catalysis is rather small. This implies either that the nitrosating species brings about substitution with an isomeric distribution only slightly different from that associated with the nitronium ion, or that the observed change in orientation is a balance of effects resulting from the operation of more than one nitrosating species, each of which effects substitution in a different isomer ratio.

Again the uncertainty about the proportion of an observed result which is due to nitration and the proportion which is due to nitrosation exists. Thus, in expt. 11 phenol was being nitrated 'above the encounter rate' and the observed isomer distribution could arise from a combination of nitration by whatever is the usual electrophile with nitration by a new, less reactive electrophile, or with nitrosation, or all three processes could be at work.

TABLE 5.7 *Orientation in the nitration of phenol*

Expt.	[PhOH] mol l^{-1}	Reagent	Additive	Concentration mol l^{-1}	Isomers (%) ortho	para	Ref.
1		HNO$_3$ (0·50 mol l^{-1}) in aq.	.	.	73, 72	.	40
2	0·45	H$_2$SO$_4$ (1·75 M), 20 °C	HNO$_2$	0·25	55, 52	.	40
3			HNO$_2$	1·00	9	.	40
4			N$_2$O$_4$	Small	44, 49	.	40
5			N$_2$O$_4$	0·03	45	.	40
6		HNO$_3$ (3·2 mol l^{-1}) in AcOH,	N$_2$O$_4$	0·27	46	.	40
7	0·6	0 °C	N$_2$O$_4$	1·8	64	.	40
8			N$_2$O$_4$	3·2	68	.	40
9			N$_2$O$_4$	3·5	70	.	40
10			N$_2$O$_4$	4·5	74	.	40
11	c. 10^{-3}	AcONO$_2$ (6·3 × 10^{-2} mol l^{-1})	.	.	43	57	11b
12	c. 10^{-3}	in Ac$_2$O, 25 °C	HNO$_2$	3 × 10^{-4}	66	33	11b
13	c. 10^{-3}	AcONO$_2$ (1·1 × 10^{-1} mol l^{-1}) in	.	.	45, 49	55, 51	11b
14	5 × 10^{-4}	Ac$_2$O, 25 °C	Urea	2 × 10^{-2}	40	60	11b
15	c. 10^{-3}	AcONO$_2$ (1·0 × 10^{-1} mol l^{-1}) in Ac$_2$O, 25 °C	HNO$_2$	4 × 10^{-4}	66	33	11b
16	c. 10^{-3}	AcONO$_2$ (2·5 × 10^{-1} mol l^{-1}) in Ac$_2$O, 25 °C	.	.	61	39	11b
17	c. 10^{-3}	AcONO$_2$ (2·2 × 10^{-1} mol l^{-1}) in	Urea	2 × 10^{-2}	64	36	11b
18	c. 10^{-3}	in Ac$_2$O, 25 °C	HNO$_2$	8 × 10^{-4}	77	23	11b

Expts. 1–3. Yields of mono-nitrophenols were 70–80 %. The yields of o-nitrophenol are subject to errors of several units %. KHSO$_4$ was present in concentration equivalent to [HNO$_3$] + [HNO$_2$].
Expts. 4–10. A small proportion of dinitration also occurred.
Expts. 11, 13, 16. Solutions prepared from pure nitric acid.
Expts. 12, 15, 18. Solutions prepared from fuming nitric acid.

Phenylboronic acid. The orientation of nitration in phenylboronic acid is very susceptible to changes in the medium (table 5.8). The high proportion of o-substitution in acetic anhydride is not attributable to a specific o-reaction, for the m:p-ratios of the last tabulated pair of results are not constant.[41] The marked change in the ratio was considered to be due to the formation in acetic anhydride of a complex, as illustrated below, which is o:p-orienting and activated as a result of the +I effect. This species need only be formed in a small concentration to overwhelm the reaction of the free acid, which is *m*-directing and deactivated because of the $-M$ effect.

TABLE 5.8   *Orientation in the nitration of phenylboronic acid*

Reagent	Isomers (%)			Ref.
	ortho	*meta*	*para*	
HNO$_3$, H$_2$SO$_4$ (1:1), -20 °C	.	70	.	42
HNO$_3$ (*d* 1·50), -15 °C	15	85	.	43
AcONO$_2$, Ac$_2$O, -15 °C	95	.	5	43
HNO$_3$, H$_2$SO$_4$ (1:1), -15 °C	22	73	5	41
AcONO$_2$, Ac$_2$O, -15 °C	63	23	14	41

Benzylic compounds. Apart from phenylboronic acid, the compounds so far considered have been very reactive ones. They are readily nitrosated, and may also be able to react with nitrating electrophiles less reactive than that operating generally in a particular nitrating system. Consequently, they are not very suitable compounds to use for the exploration of the characteristics of these systems. For this reason, experiments which have been carried out on a number of moderately reactive benzylic compounds are especially interesting. Results are given in table 5.9. We are especially concerned here with the distribution of isomers; other aspects of the nitration of benzylic compounds will be discussed later (§9.1.1). The nitrations of the first five compounds listed in table 5.9 lead to isomer distributions which change little or not at all with the reagent; this circumstance has been encountered with a number of other compounds (see above). In contrast, the *o*:*p*-ratio for nitration of methyl phenethyl ether with acetyl nitrate is considerably higher than the value for nitration with mixed acid or with nitric acid.

In accounting for such a change three possible circumstances have to be considered: first, the active electrophile may vary from one nitrating agent to the other; second, the character of the aromatic compound may be modified by a change in the nitrating system; and third, both of these circumstances may occur. In their first discussion of the problem Norman and Radda[39] argued strongly for the first of these circumstances. Since the *m*:*p*-ratios for the nitrations of the ether were

TABLE 5.9   *Orientation in the nitration of some benzylic compounds*

		Isomers (%)			
Compound	Reagent	ortho	meta	para	Ref.
1 Benzyl chloride*	AcONO$_2$, 25 °C	33·6	13·9	52·5	20, 44
2	HNO$_3$, H$_2$SO$_4$ (1:1), 25 °C	34·4	14·1	51·5	20, 44
3 Benzyl cyanide	AcONO$_2$, 25 °C	24·4	20·1	55·5	20, 44
4	HNO$_3$, H$_2$SO$_4$ (1:1), 25 °C	22·0	20·7	57·3	20, 44
5 Ethyl phenyl-acetate*	AcONO$_2$, 25 °C	54·3	13·1	32·6	20, 44
6	HNO$_3$, H$_2$SO$_4$ (1:1), 25 °C	41·7	24·6	33·7	20, 44
7 Phenylnitro-methane	AcONO$_2$, 25 °C	22·5	54·7	22·8	20, 44
8	HNO$_3$, H$_2$SO$_4$ (1:1), 25 °C	22·2	53·1	24·7	20, 44
9 Toluene	Ac$_2$O, 25 °C	56·1	2·5	41·4	20, 44
10	HNO$_3$, H$_2$SO$_4$ (1:1), 25 °C	56·0	2·4	41·6	20, 44
11 Benzyl methyl ether	AcONO$_2$, 25 °C	51·3	6·8	41·9	20, 44
12	Fuming HNO$_3$ (d 1·5), 25 °C	38·5	12·8	48·7	39
13	HNO$_3$, H$_2$SO$_4$ (1:1), 25 °C	28·6	18·1	53·3	20, 44
14 Methyl phenethyl ether	AcONO$_2$, 0 °C	64·4	3·6	32·0	39
15	AcONO$_2$, 25 °C	62·3	3·7	34·0	39, 44
16	AcONO$_2$ (1·0 mol l^{-1}), Ac$_2$O, 25 °C	60	5	33	11b
17	AcONO$_2$ (1·0 mol l^{-1}), Ac$_2$O, 25 °C +NaNO$_3$ (0·02 mol l^{-1})	57	5	36	11b
18	AcONO$_2$, MeCN, 0 °C	66·0	4·2	29·8	39
19	HNO$_3$ (1·0 mol l^{-1}), Ac$_2$O, 25 °C	59	5	34	11b
20	HNO$_3$, MeNO$_2$, 25 °C	41·2	3·0	55·8	39
21	Fuming HNO$_3$ (d 1·5), 25 °C	40·2	6·6	53·2	39
22	HNO$_3$, H$_2$SO$_4$ (1:1), 0 °C	28·9	8·7	62·4	39
23	HNO$_3$, H$_2$SO$_4$ (1:1), 25 °C	31·6	9·4	59·0	39,44
24	HNO$_3$ (0·015 mol l^{-1}), 65% H$_2$SO$_4$, 25 °C	34	9	57	11b
25	Benzoyl nitrate, MeCN	65·7	3·6	30·7	39
26	p-Nitrobenzoyl nitrate, MeCN	64·9	4·7	30·4	39
27	N$_2$O$_5$, MeCN, 0 °C	68·8	3·3	27·9	39
28 Methyl 3-phen-propyl ether	AcONO$_2$, 0 °C	43·0	4·0	53·0	39
29	Fuming HNO$_3$ (d 1·5), 0 °C	44·2	3·8	52·0	39
30	HNO$_3$, H$_2$SO$_4$	37·3	5·9	56·8	44

Expts. 1, 3, 5, 7, 9, 11. Descriptions of these experiments are not always explicit, but the reagent was prepared from fuming nitric acid (d 1·5, 0·009 mol) and acetic anhydride (0·01 mol); a small quantity of urea was added before nitration.

Expts. 2, 4, 6, 8, 10, 13, 30. Fuming nitric acid (d 1·5) was used in preparing the reagent.

Expts. 14, 15, 28. The reagent was prepared from fuming nitric acid (d 1·5, 0·005 mol) and acetic anhydride (0·005 mol) at 0 °C.

Expts. 16, 17. Pure nitric acid was used. In expt. 16 the reaction was of the first order in the concentration of the aromatic, and of half-life 1–1·5 minutes (similar to that of toluene under the same conditions). In expt. 17 the sodium nitrate slowed the reaction (half-life c. 60 min). About 2% of an acetoxylated product was formed (table 5·4).

essentially the same no matter whether acetyl nitrate or mineral acid was used, the change in $o:p$-ratios could not be due to protonation of the ether in the mineral acid; protonation would be expected to produce a higher $m:p$-ratio. In mixed acid, as against nitric acid, the slightly greater proportion of m-nitration might indicate a small degree of protonation. The variation in $o:p$-ratios cannot result simply from diminished reactivity of the o-position in acid solutions.

The behaviour of compounds like toluene suggested that the nitronium ion may be the nitrating species common to mineral acid solutions and acetyl nitrate, and there must be another species present in acetyl nitrate capable only of nitrating the o-position in the ether but not of nitrating the p-position, or of nitrating any of the positions in toluene. Dinitrogen pentoxide was suggested to be this species because of its known presence in the system, and because other systems which are believed to produce it, such as benzoyl and p-nitrobenzoyl nitrate, as well as dinitrogen pentoxide itself in acetonitrile, gave closely similar high $o:p$-ratios. In sum, 'the acyl nitrate, itself not reactive enough to bring about nitration of the aromatic compound, gives rise to dinitrogen pentoxide (and acyl anhydride) which reacts in two ways. It undergoes slow heterolysis to the nitronium ion which reacts with the aromatic compound to give o-, m-, and p-nitro-derivatives in the same proportions as when nitric acid is the reagent; and at the same time there is an additional mode of nitration at the o-position, dependent on the presence of the oxygen atom of the ether, and brought about by dinitrogen pentoxide itself.'[39] The interaction envisaged is shown overleaf.

In a later paper Knowles and Norman[44] compared more fully nitrations of benzylic compounds in acetyl nitrate and in mixed acid (table 5.9), and interpreted the results in terms of three factors: nitronium ion nitration in both media; some degree of protonation of the oxygen

---

Table 5.9 (*cont.*)

Expts. 18, 25, 26. Silver nitrate (0·005 mol) and the aromatic compound (0·01 mol) were dissolved in acetonitrile (1 ml) and a solution of the acyl chloride (0·005 mol) in acetonitrile (0·4 ml) was added.

Expt. 19. The aromatic compound was added to a freshly prepared solution of nitric acid in acetic anhydride. The reaction was very fast (< 1 min.) About 2 % of an acetoxylated product was formed (table 5.4).

Expt. 20. The results are not as accurate as for other expts. because of the small degree of nitration achieved.

Expts. 21. Heterogeneous reaction.

Expt. 27. A solution of dinitrogen pentoxide (0·005 mol) in acetonitrile (0·4 ml) was added slowly to the aromatic compound (0·01 mol) in acetonitrile (1 ml).

* Slightly different results were reported by earlier workers.[16e]

atom of the ether (whether involving complete proton transfer or not); and the presence of dinitrogen pentoxide in acetyl nitrate solutions, permitting specific *o*-interaction with the oxygen atom.

Evidence for the influence of protonation was convincingly adduced from the trend of the quantity $m:p$ (mixed acid)/$m:p$ (acetyl nitrate) in the series $Ph.CH_2.OMe$, $Ph.(CH_2)_2.OMe$, $Ph.(CH_2)_3.OMe$, but it was argued that protonation in mixed acid cannot explain the change in *o*:*p*-ratios with change of nitrating conditions. Thus, it was supposed that by analogy with $Ph.CH_2\overset{+}{N}Me_3$ nitration of the conjugate acid of benzyl methyl ether would give 80 % of the m-nitro-compound; then from expts. 11 and 13 (table 5.9) in mixed acid 15 % of the reaction involves the conjugate acid (the original paper says about 12 %). Then, even with the extreme supposition that nitration of the conjugate acid gives no *o*-nitro-derivatives, the proportion of *o*-nitro-derivative produced in acetyl nitrate, arising wholly from the free base, would be 33 %. It is, in fact, 51·3 %. Further, the value of the ratio $o : p$ (acetyl nitrate)/$o:p$ (mixed acid) along the series

$Ph.CH_2.OMe$, $Ph.(CH_2)_2.OMe$, $Ph.(CH_2)_3.OMe$ (2·3, 3·4, 1·3),

does not decrease steadily, but goes through a maximum. These two circumstances point to a specific *o*-interaction in nitrations of the ethers with acetyl nitrate which is important with benzyl methyl ether, more important with methyl phenethyl ether, and not important with 'methyl phenpropyl ether. This interaction is the reaction with dinitrogen pentoxide already mentioned, and the variation in its importance is thought to be due to the different sizes of the rings formed in the transition states from the different ethers.

In considering these results and the explanation suggested for them, it may first be noticed that some of the subsidiary points in the argument might now need to be expressed differently; current views about the inductive effect (§9.12) might make the question of the $m:p$-ratio appear differently, but could not much alter the situation, for even if changes in this quantity were to be expected they would be of a minor order.

The crucial questions are really three: does any one of the ethers really stand out from the others as having a particularly high $o:p$-ratio; does such a high $o:p$-ratio require a specific o-interaction between the ether and the electrophile to account for it; does the identification of a specific o-interaction require the intervention of dinitrogen pentoxide?

The arguments of Norman and his co-workers seem to give affirmative answers to the first and second of these questions, but it is doubtful if the available data further require such an answer for the third question. It can be argued[11b] that the crucial comparison made between the behaviour of benzyltrimethylammonium ion and protonated benzyl methyl ether is invalid, and that it is possible to interpret the results in terms of nitration by the nitronium ion, modified by protonation of the oxygen atom of the ether; a case for the possible involvement of the nitronium ion in specific interaction leading to o-substitution has been made.[29, 45]

Such arguments are based on the assumption that the nitronium ion is the nitrating agent in all of the media under consideration; as regards nitration with acetyl nitrate, they certainly do not prove the efficacy of the nitronium ion unless the participation of the latter can be shown to be also consistent with the kinetic evidence.

5.3.5 *The mechanisms of nitration with acetyl nitrate in acetic anhydride*

By analogy with the mechanisms of nitration in other media, and from a knowledge of the composition of solutions of acetyl nitrate in acetic anhydride, the following may be considered possible nitrating species in these solutions:

$$HNO_3, \quad H_2NO_3^+, \quad AcONO_2, \quad AcONO_2H^+, \quad NO_2^+, \quad N_2O_5.$$

The observation of nitration at a rate independent of the concentration and nature of the aromatic excludes $AcONO_2$ as the reactive species. The fact that zeroth-order rates in these solutions are so much faster than in solutions of nitric acid in inert organic solvents, and the fact that HNO_3 and $H_2NO_3^+$ are ineffective in nitration even when they are present in fairly large concentrations, excludes the operation of either of these species in solutions of acetyl nitrate in acetic anhydride.

The facts, in particular the dependence of first-order rate upon the concentration of acetyl nitrate (Appendix),[11c] could not be accounted for if protonated acetyl nitrate were the reagent. The same objections apply to the free nitronium ion. It might be possible to devise a means of generating dinitrogen pentoxide which would account for the facts of zeroth- and first-order nitration, but the participation of this reagent could not be reconciled with the anticatalysis by nitrate of first-order nitration.

Another reagent which must be considered is the ion pair $AcONO_2H^+$ NO_3^-, the species favoured by Fischer, Read and Vaughan.[14b] Its participation would make it possible to account for the dependence of rate of zeroth-order nitration upon the concentration of acetyl nitrate and acetic acid, and would lead to the prediction of similar dependencies in first-order nitration. It would not, however (*pace* Fischer, Read and Vaughan[14b]), explain the anticatalytic effect of added nitrate.

Thus, strong arguments against all of the obvious nitrating species acting alone can be found. However, as has been pointed out, the extent to which ions require solvation by nitric acid molecules in this medium is unknown, and such solvation would influence the apparent order with respect to the stoichiometric nitric acid.[45] The possibility also exists that more than one mechanism of nitration, excluding nitrosation, is operative.

The fact that nitration with acetyl nitrate is sometimes accompanied by acetoxylation has been mentioned (§5.3.3). In proposing the ion pair $AcONO_2H^+$ NO_3^- as the nitrating agent, Fischer, Read and Vaughan[14b] also suggested that it was responsible for the acetoxylation, which was regarded as an electrophilic substitution.

If acetoxylation were a conventional electrophilic substitution it is hard to understand why it is not more generally observed in nitration in acetic anhydride. The acetoxylating species is supposed to be very much more selective than the nitrating species, and therefore compared with the situation in (say) toluene in which the ratio of acetoxylation to nitration is small, the introduction of activating substituents into the aromatic nucleus should lead to an increase in the importance of acetoxylation relative to nitration. This is, in fact, observed in the limited range of the alkylbenzenes, although the apparently severe steric requirement of the acetoxylation species is a complicating feature. The failure to observe acetoxylation in the reactions of compounds more reactive than *m*-xylene has been attributed[14] to the incursion of another mechanism for nitration, first order in the concentration of aromatic, in which the species responsible for nitration does not also acetoxylate. However, later results show that no such mechanism intervenes, for the rates of reaction of mesitylene, anisole, *o*- and *m*-xylene were all found to be independent on the concentration of aromatic, provided that this was greater than about $0 \cdot 1$ mol l^{-1}. Therefore there is a continuity of mechanism up to the level of reactivity of anisole and mesitylene, and evidence has been presented that this is a limiting level of reactivity in this medium. Ridd[45] has suggested that the acetoxylations are addition-elimination reactions in which acetate adds to the σ-complex formed from the aromatic compound and nitronium ion. It remains to be seen if this sort of mechanism can account for the narrow conditions of aromatic structure which are evidently characteristic of acetoxylation.

REFERENCES

1. Francis, F. E. (1906). *J. chem. Soc.* **89**, 1; *Ber. dt. chem. Ges.* **39**, 3798.
2. Gold, V., Hughes, E. D. & Ingold, C. K. (1950). *J. chem. Soc.* p. 2467.
3. Vandoni, R. & Viala, P. (1945). *Mém. Services chim. État* **32**, 80.
4. Chédin, J. & Fénéant, S. (1949). *C. r. hebd. Séanc. Acad. Sci., Paris* **229**, 115.
5. Marcus, R. A. & Frescoe, J. M. (1957). *J. chem. Phys.* **27**, 564.
6. Mal'kova, T. V. (1954). *Zh. obshch. Khim.* **24**, 1157.
7. Lloyd, L. & Wyatt, P. A. H. (1957). *J. chem. Soc.* p. 4268.
8. Pictet, A. & Khotinsky, E. (1907). *C. r. hebd. Séanc. Acad. Sci., Paris* **144**, 210.
9. Bordwell, F. G. & Garbisch, E. W. (a) (1960). *J. Am. chem. Soc.* **82**, 3588.
   (b) (1962). *J. org. Chem.* **27**, 2322, 3049.
   (c) (1963). *J. org. Chem.* **28**, 1765.
10. Mantsch, O., Bodor, N. & Hodorsan, F. (1968). *Rev. chim. (Roumania)* **13**, 1435.
11. (a) Hoggett, J. G., Moodie, R. B. & Schofield, K. (1969). *Chem. Comm.* p. 605.
    (b) Hoggett, J. G. (1969). Ph. D. thesis, University of Exeter.
    (c) Hartshorn, S. R., unpublished results.
    (d) Thompson, M. J. unpublished results.
12. Chattaway, F. D. (1910). *J. chem. Soc.* p. 2099.
13. Cohen, F. H. & Wibaut, J. P. (1935). *Recl Trav. chim. Pays-Bas Belg.* **54**, 409.
14. (a) Fischer, A., Packer, J., Vaughan, J. & Wright, G. J. (1964). *J. chem. Soc.* p. 3687.
    (b) Fischer, A., Read, A. J. & Vaughan, J. (1964). *J. chem. Soc.* p. 3691.
    (c) Fischer, A., Vaughan, J. & Wright, G. J. (1967). *J. chem. Soc. B*, p. 368.
15. Wibaut, J. P. (1915). *Recl Trav. chim. Pays-Bas Belg.* **34**, 241.

References

16. Ingold, C. K. (a) with Shaw, F. R. (1927). *J. chem. Soc.* p. 2918; (b) with Lapworth, A., Rothstein, E. & Ward, D. (1931). *J. chem. Soc.* p.1959; (c) with Smith, M. S. (1938). *J. chem. Soc.* p. 905; (d) with Bird, M. L. (1938). *J. chem Soc.* p. 918; (e) with Shaw, F. R. (1949). *J. chem. Soc.* p. 575.
17. Stock, L. M. (1961). *J. org. Chem.* **36**, 4120.
18. Bordwell, F. G. & Rohde, K. (1948). *J. Am. chem. Soc.* **70**, 1191.
19. Knowles, J. R., Norman, R. O. C. & Radda, G. K. (1960). *J. chem. Soc.* 4885.
20. Knowles, J. R. & Norman, R. O. C. (1962). *J. chem. Soc.* p. 2938.
21. Speier, J. L. (1953). *J. Am. chem. Soc.* **75**, 2930.
22. Cohn, H., Hughes, E. D., Jones, M. H. & Peeling, M. G. (1952). *Nature, Lond.* **169**, 291.
23. Dewar, M. J. S. (a) with Mole, T. (1956). *J. chem. Soc.* 1441; (b) with Mole, T., Urch, D. S. & Warford, E. W. T. (1956). *J. chem. Soc.* p. 3572; (c) with Mole, T. & Warford, E. W. T. (1956). *J. chem. Soc.* p. 3576; (d) with Mole, T. & Warford, E. W. T. (1956). *J. chem. Soc.* p. 3581; (e) with Logan, R. H. (1968). *J. Am. chem. Soc.* **90**, 1924.
24. Dewar, M. J. S. & Urch, D. S. (a) (1957). *J. chem. Soc.* p. 345. (b) (1958). *J. chem. Soc.* p. 3079.
25. Streitwieser, A. & Fahey, R. C. (1962). *J. org. Chem.* **27**, 2352.
26. Taylor, R. (1966). *J. chem. Soc.* B, p. 727.
27. Cohen, F. H. (1928). *Proc. Acad. Sci., Amsterdam* **31**, 692.
28. Paul, M. A. (1958). *J. Am. chem. Soc.* **80**, 5329.
29. de la Mare, P. B. D. & Ridd, J. H. (1959). *Aromatic Substitution: Nitration and Halogenation*, p. 76. London: Butterworths.
30. Halvarson, K. & Melander, L. (1957). *Ark. Kemi.* **11**, 77.
31. Paul, M. A. (1958). *J. Am. chem. Soc.* **80**, 5332.
32. Sparks, A. K. (1966). *J. org. Chem.* **31**, 2299.
33. Griffiths, P. H., Walkey, W. A., & Watson, H. B. (1934). *J. chem. Soc.* p. 631.
34. Bunton, C. A., Minkoff, G. J. & Reed, R. I. (1947). *J. chem. Soc.* p. 1416.
35. Coombes, R. G., Moodie, R. B., & Schofield, K. (1968) *J. chem. Soc.* B, p. 800.
36. Arnall, F. & Lewis, T. (1929). *J. Soc. chem. Ind.* **48**, 159 T.
37. Lynch, B. M., Chen, C. M. & Wingfield, Y.-Y. (1969). *Can. J. Chem.* **46**, 1141.
38. Holleman, A. F. (1925). *Chem. Rev.* **1**, 187.
39. Norman, R. O. C. & Radda, G. K. (1961). *J. chem. Soc.* p. 3030.
40. Bunton, C. A., Hughes, E. D., Ingold, C. K., Jacobs, D. I. H., Jones, M. H., Minkoff, G. J. & Reed, R. I. (1950). *J. chem. Soc.* p. 2628.
41. Harvey, D. R. & Norman, R. O. C. (1962). *J. chem. Soc.* p. 3822.
42. Ainley, A. D. & Challenger, F. (1930). *J. chem. Soc.* p. 2171.
43. Seamon, W. & Johnson, J. R. (1931). *J. Am. chem. Soc.* **53**, 711.
44. Knowles, J. R. & Norman, R. O. C. (1961). *J. chem. Soc.* p. 3888.
45. Ridd, J. H. (1966). *Studies on Chemical Structure and Reactivity*, ed. J. H. Ridd. London: Methuen.

6 The process of substitution

6.1 INTRODUCTION

In earlier chapters we have been concerned with the identification of the effective electrophile in nitrations carried out under various conditions. We have seen that very commonly the nitronium ion is the electrophile, though dinitrogen pentoxide seems capable of assuming this role. We now consider how the electrophile, specifically the nitronium ion, reacts with the aromatic compound to cause nitration.

For electrophilic substitutions in general, and leaving aside theories which have only historical interest,[1] two general processes have to be considered. In the first, the S_E3 process, a transition state is involved which is formed from the aromatic compound, the electrophile (E^+), and the base (B) needed to remove the proton:

$$ArH + E^+ + B \xrightarrow{k_3} \{(E\ldots Ar\ldots H\ldots B)^+\}^* \rightarrow ArE + BH^+.$$

For such a process the following rate equation holds:

$$\text{Rate} = k_3 \, [ArH] \, [E^+] \, [B].$$

In the second, the S_E2 process, there are two stages:

$$ArH + E^+ \underset{k_{-1}}{\overset{k_1}{\rightleftharpoons}} E - \overset{+}{Ar}\!\!-\!H,$$

$$E\!-\!\overset{+}{Ar}\!\!-\!H + B \xrightarrow{k_2} ArE + BH^+,$$

and two limiting forms are recognisable. In one there is formation of a bond between the electrophile and the aromatic nucleus in a rate-determining step; and, in the other, formation of the bond between the electrophile and the aromatic nucleus is followed by rate-determining loss of the proton. In the latter circumstance an equilibrium concentration of an intermediate, $E\!-\!\overset{+}{Ar}\!\!-\!H$, will be formed.

The rate equation for the S_E2 form is, from the steady-state approximation, as follows:

$$\text{rate} = \frac{k_1 \, k_2 \, [B] \, [ArH] \, [E^+]}{k_{-1} + k_2 \, [B]}.$$

When $k_2[B] \ll k_{-1}$ the S_E2 process will show a linear dependence of rate upon the concentration of base. When $k_2[B] \gg k_{-1}$ a reaction occurring by this mechanism would not be subject to base catalysis. Between these extremes a range of behaviour is possible.

For electrophilic substitutions in general, some form of the S_E2 mechanism is now believed to operate.[2, 3] We can now review the evidence concerning the particular case of nitration.

6.2 NITRATIONS WITH NITRONIUM IONS: THE GENERAL CASE

6.2.1 *Evidence from solvent effects*

Hughes, Ingold and Reed[4] discussed the relative merits of the S_E2 and S_E3 schemes as mechanisms for nitration by considering the properties of acetic acid, nitromethane, nitric acid and sulphuric acid as media for the reaction. The facts have already been discussed (§§2.2.3, 2.2.4, 2.3.2, 2.4.2, 2.4.3, 3.2).

The process of nitration can be expressed in the following compressed form:

$$HNO_3 \underset{b}{\overset{a}{\rightleftharpoons}} NO_2^+ \overset{c}{\longrightarrow} ArNO_2.$$

All of the reactions leading to the generation of the nitronium ion are summarized in a, processes reversing this generation in b, and all the processes involved in the attack by nitronium ion upon the aromatic and expulsion of the proton in c.

In the cases of nitration in the organic solvents, the process a leads to the formation of two ions and a neutral molecule from two neutral molecules (§3.2.4) and b to the reverse of this. If c summarizes the S_E2 process, the formation of E–Ar–H being rate-determining, then in it ionic charge is neither formed nor destroyed. In these circumstances, a should be accelerated by increased polarity of the solvent, b should be retarded, and, to a first approximation, c should be unaffected. Zeroth-order rates, which depend only on a, should therefore be increased in more polar solvents, as should first-order rates because of the consequent increase in nitronium ion concentration. The kinetic order of nitration depends upon b and c; whatever decreases the rate of b relative to that of c will tend to move the kinetic order from unity to zero, and vice versa. Changes in solvent polarity thus affect kinetic order in the expected way (§3.2.1).

The S_E3 mechanism does not satisfactorily account for the kinetic observations, as can be seen when the nature of the base (B) in nitrations is recalled; it would most likely be a nitrate or bisulphate anion.

The argument for the S_E2 process, when the transition from acetic acid as solvent to nitric acid as solvent is considered, is less direct, for because of the experimental need to use less reactive compounds, zeroth-order nitration has not been observed in nitric acid. It can be estimated, however, that a substance such as nitrobenzene would react about 10^5 faster in first-order nitration in nitric acid than in a solution of nitric acid (7 mol l^{-1}) in acetic acid. Such a large increase is understandable in terms of the S_E2 mechanism, but not otherwise.

As regards sulphuric acid, there is here again an increase in polarity and an increase in rates of nitration when comparison is made with other solvents in the series. This gross fact would be difficult to reconcile with any mechanism, such as the S_E3 one, which contains an essential forward step which would be retarded by increased polarity of the solvent.

In originally considering the S_E3 mechanism, involving base catalysis, Bennett, Brand, James, Saunders and Williams[5] were trying to account for the small increase in nitrating power which accompanies the addition of water, up to about 10%, to sulphuric acid. The dilution increases the concentration of the bisulphate ion, which was believed to be the base involved (along with molecular sulphuric acid itself). The correct explanation of the effect has already been given (§2.3.2).

Thus, solvent effects are consistent with a S_E2 mechanism in which the rate-determining step, leading to the intermediate, E–Ar–H, does not involve a base, and in which the proton is lost in a subsequent step of no kinetic importance.

In the general context of the effect of medium upon rate, the data given in table 4.4 are of interest.

6.2.2 *Evidence from isotope effects*

If, in the transition state of the rate-determining step of a substitution reaction, the bond between the hydrogen atom and the ring which is finally broken is significantly stretched, then the corresponding replacements of deuterium or tritium should show primary kinetic isotope effects. The difference in zero-point energy between carbon–protium and carbon–deuterium or carbon–tritium bonds should be greater in the ground state than in the transition state, leading to considerably slower replacement of deuterium or tritium than of protium.[6]

At one time a form of S_E2 mechanism was favoured for electrophilic substitution in which in the transition state bonding between carbon and the electrophile and severance of the proton had proceeded to the same degree, without interference with the aromatic system. Such a transition state had the advantage of retaining the stability of the aromatic ring arising from delocalisation.[7] With the inclusion of the base needed to remove the proton this would serve as a model for the transition state of a S_E3 reaction. The degree to which such a mechanism would exhibit a primary kinetic isotope effect would depend upon the degree to which the carbon–hydrogen bond had been stretched in the transition state. Indeed, although it seems probable that such a mechanism requires a primary kinetic isotope effect, the absence of this effect would not conclusively eliminate the mechanism.[8]

With the S_E2 mechanism several circumstances are possible. When $k_2[B] \gg k_{-1}$ the loss of the proton would be kinetically unimportant, and the transition state in the rate-determining step, leading to $E\overset{+}{-Ar}-H$, could be one in which the carbon–hydrogen bond was not much modified from its original condition. No kinetic isotope effect would be expected. If, however, $k_2[B] \ll k_{-1}$, the loss of the proton would be important, and a primary kinetic isotope effect would be observed. As with the phenomenon of base catalysis, between these extremes a range of possibilities exists.

It is clear, then, that the measurement of primary kinetic isotope effects will not give a wholly unambiguous clue to mechanism in the absence of other evidence. Nevertheless, the absence of a kinetic isotope effect is most easily understood in terms of the S_E2 mechanism leading to the formation of an intermediate $E\overset{+}{-Ar}-H$.

For electrophilic substitutions in general, a range of behaviour has been observed, all of which can be accommodated by the two-stage mechanism, leading to the strong presumption that this two-stage mechanism is general.[2, 3] The kind of supporting evidence which is especially telling is the observation of base-catalysis in some form.[3] Nitrations are carried out in circumstances which do not favour the ready examination of this point, and the supposition that certain phenomena could be ascribed to it has already been shown to be ill-founded (§6.2.1).

Melander first sought for a kinetic isotope effect in aromatic nitration; he nitrated tritiobenzene, and several other compounds, in mixed acid and found the tritium to be replaced at the same rate as protium (table 6.1).[6b] Whilst the result shows only that the hydrogen is not appreciably loosened in the transition state of the rate-determining step, it is most easily understood in terms of the S_E2 mechanism with

TABLE 6.1   *Hydrogen isotope effects in nitrations*

Compound	Conditions	k_H/k_D	k_H/k_T	Ref.
Anisole-o-T	Benzoyl chloride–silver nitrate –acetonitrile, 0 °C	.	1	9
Anthracene-9-D	$NO_2^+BF_4^-$–sulpholan, 30 °C	2·6±0·3	.	33
	$NO_2^+BF_4^-$–acetonitrile, 0 °C	6·1±0·6	.	33
Benzene-T	1:2 nitric acid (d 1·40) –sulphuric acid (96%), 77 °C	.	< 1·19	6b
Benzene-D	Nitric acid–sulphuric acid	1	.	10
Benzene-D₆	$NO_2^+BF_4^-$–sulpholan, 25 °C	0·89	.	11a
1,3,5-tri-t-Butyl benzene-T	Nitric acid (d 1·5)–acetic acid acid–acetic anhydride, c. 25 °C	.	1	12a
1,3,5-tri-t-Butyl benzene-D	As above	1·0	.	12b
1,3,5-tri-t-Butyl-2-fluorobenzene-D₂	Nitric acid (90%)–sulphuric acid (96%)–nitromethane, 40°C	2·3	.	12b
1,3,5-tri-t-Butyl-2-methylbenzene-D₂	As above	3·7	.	12b
1,3,5-tri-t-Butyl-2-nitrobenzene-D₂	As above	3·0	.	12b
Bromobenzene-T	1:1 nitric acid (d 1·42)-oleum (20–30% SO_3), 77 °C	.	< 1·28	6b
p-Bromonitrobenzene-T	As above	.	< 1·23	6b
Fluorobenzene-p-D	$NO_2^+BF_4^-$–sulpholan, 25 °C	0·82	.	11b
2,6-Lutidine-4-D 1-oxide	Sulphuric acid (87·9%), 80 °C (kinetic)	1	.	13
Naphthalene-α-T	Nitric acid (d 1·40)+2% (v/v) of water, 77 °C	.	<1·30	6b
Naphthalene-1,4-D₂	$NO_2^+BF_4^-$–sulpholan, 30 °C	1·15±0·05	.	33
	$NO_2^+BF_4^-$–acetonitrile, 0 °C	1·08±0·05	.	33
Nitrobenzene-T	1:2 nitric acid (d 1·40) –sulphuric acid (96%), 77 °C	.	< 1·35	6b
Nitrobenzene-D	Nitric acid–sulphuric acid	1	.	10
Nitrobenzene-D₅	97·4% and 86·7% sulphuric acid, 25 °C (kinetic)	1	.	14
Thiophen-2-T	Benzoyl nitrate–acetonitrile, −2 °C	.	0·88	15
Toluene-α-T,-o-T, -m-T and -p-T	1:2 nitric acid (d 1·40)-sulphuric acid (96%), 77 °C	.	< 1·18	6b
Toluene-o-T	Benzoyl chloride–silver nitrate-acetonitrile, 0 °C	.	1	9
Toluene-α-T, -m-T and -p-T	Nitric acid (d 1·42)–sulphuric acid 0 °C	.	1	16
Toluene-α-D₃	Nitric acid (d 1·51)–nitromethane 20° C	1·04±0·07	.	17
Toluene-o-D, -m-D, and -p-D	As above	.	1·00	17
Toluene-α-D₃ and -α-T	Nitric acid (80% vol.), 25 °C	1·002	1·003	18
Toluene-D₈	$NO_2^+BF_4^-$–sulpholan, 25 °C	0·85	.	11a

a relatively stable intermediate. Otherwise it would, in Melander's words, 'be hard to understand how, in the imaginary reversed process, it were possible to bring a hydrogen nucleus up to its normal bond distance from the carbon without reaching the potential energy maximum (from the other side). It seems natural to assume that the intermediate corresponds to a real minimum in the energy surface, and that saddle points on each side correspond to the losses of NO_2^+ or H^+. The first saddle point is at the highest level.'

One way in which the step of the reaction in which the proton is lost might be slowed down, and perhaps made kinetically important (with $k_2[B] \gg k_{-1}$), would be to carry out nitration at high acidities. Nitration of pentadeuteronitrobenzene in 97·4% sulphuric acid failed to reveal such an effect.[14] In fact, nitrations under a variety of conditions fail to show a kinetic isotope effect.

The picture of the process of substitution by the nitronium ion emerging from the facts discussed above is that of a two-stage process, the first step in which is rate-determining and which leads to a relatively stable intermediate. In the second step, which is relatively fast, the proton is lost. The transition state leading to the relatively stable intermediate is so constructed that in it the carbon–hydrogen bond which is finally broken is but little changed from its original condition.

Beyond this we know little or nothing about the structure of the transition state, though there are arguments to suggest that in it the aromatic ring retains most of its delocalisation energy (§7.2.3). In particular, we do not know how the nitronium ion is oriented in the transition state. The nitronium ion presumably approaches the carbon atom being attacked in a plane at right angles to that of the aromatic ring,[19] and it has been suggested that in the transition state it has become slightly bent, with its two oxygen atoms situated over the two positions in the ring adjacent to the one being substituted; this might account for any greater sensitivity of nitration to steric hindrance than of bromination with the brominium cation[20] (see §6.2).

6.2.3 *The nature of intermediates in nitrations*

The structure of the intermediate was represented by Melander[6b] and by Ingold[21] as shown in (I) below, the positive charge of the nitronium ion now being carried by the ring. This is conveniently written as (II), though the positive charge cannot be regarded as being uniformly distributed amongst five of the carbon atoms of the ring [see (III)].

There is evidence for the existence of structures of this kind, and for their importance in electrophilic substitution in general, and in nitration in particular. Because of the way in which the electrophile is attached to the ring they are called σ-complexes.

(I)   (II)   (III)

(IV)   (V)

With hydrogen chloride and aluminium chloride,[22a] or hydrogen bromide and aluminium bromide,[22b, 23] homologues of benzene give complexes which are coloured, electrically conducting, and are formed at measurable rates at low temperatures. In them deuterium and protium undergo rapid exchange. The use of hydrogen fluoride and boron trifluoride gives complexes with similar properties.[24] There can be no doubt that these various products are σ-complexes with structures of the type (IV).

Cations like that present in (IV) exist in solutions of aromatic hydrocarbons in trifluoroacetic acid containing boron trifluoride,[25] and in liquid hydrogen fluoride containing boron trifluoride.[26] Sulphuric acid is able to protonate anthracene at a *meso*-position to give a similar cation.[27]

The relative basicities of aromatic hydrocarbons, as represented by the equilibrium constants for their protonation in mixtures of hydrogen fluoride and boron trifluoride, have been measured.[24a] The effects of substituents upon these basicities resemble their effects upon the rates of electrophilic substitutions; a linear relationship exists between the logarithms of the relative basicities and the logarithms of the relative rate constants for various substitutions, such as chlorination[22c, 28-9] and protodedeuteration in a mixture of trifluoroacetic and sulphuric acid,[30a] into the homologues of benzene. Similar relationships exist for the deuterodeprotonation, chlorination in acetic acid and nitration in acetic anhydride of polynuclear hydrocarbons.[30b] (Comments upon the data for nitration are made later (§7.3.1).) The implication of these relationships is that the values of the Gibbs function of the σ-complexes are determined by factors which also control the values of the Gibbs function of the transition states in the rate-determining steps of the substitution reactions and resemble each other; the σ-complexes resemble the transition state more than either resembles the reactants, and the σ-complexes are better models for the transition states than are the reactants.[8]

The σ-complexes (IV) are thus the intermediates corresponding to the substitution process of hydrogen exchange. Those for some other substitutions have also been isolated; in particular, benzylidyne trifluoride reacts with nitryl fluoride and boron trifluoride at $-100\ ^\circ$C to give a yellow complex. Above $-50\ ^\circ$C the latter decomposes to hydrogen fluoride, boron trifluoride, and an almost quantitative yield of *m*-nitrobenzylidyne trifluoride.[31] The latter is the normal product of nitrating benzylidyne trifluoride, and the complex is formulated as (v).

Pentamethylbenzene[32a] and anthracene[33] react very rapidly with nitronium tetrafluoroborate in sulpholan to give σ-complexes, which decompose slowly (see below), and durene behaves similarly with nitronium hexafluorophosphate in acetonitrile.[32b]

6.3 NITRATIONS WITH NITRONIUM IONS: SPECIAL CASES

As outlined above, the process of substitution by the nitronium ion is satisfactorily described by an S_E2 mechanism in which $k_2[B] \gg k_{-1}$. In certain circumstances the process could be changed so that this condition did not hold, and the step in which the proton is lost could become kinetically important. One such circumstance is that in which the hydrogen atom being replaced is situated between bulky substituents; steric hindrance would then make it difficult for the nitro group to move from its position in the intermediate complex to that between the bulky substituents; k_2 would be diminished, and a kinetic isotope effect might appear. It is for this reason that 1,3,5-tri-*t*-butylbenzene and its derivatives are interesting (table 6.1); whilst the hydrocarbon undergoes nitration without a kinetic isotope effect this is not the case with its fluoro, nitro, and methyl derivatives. In the series of compounds 1,3,5-tri-*t*-butylbenzene and its fluoro, nitro and methyl derivatives the kinetic isotope effects (k_H/k_D = 1·0, 2·3, 3·0 and 3·7) are not related to the overall rates of nitration (approximate relative rates: 1, 0·1, 0·001 and 10) and so not to the electronic characters of the extra substituents. They are, however, roughly related to the sizes of the substituents. Accordingly, the data are interpreted[12b] as indicating the operation of steric repulsions which differentially retard the second step of the reaction as compared with the reversal of the first step. In these respects nitration is less demanding than bromination with Br+,[12b] which shows a kinetic isotope effect with 1,3,5-tri-*t*-butylbenzene itself. This could mean either that in bromination reversal of the first step was more sensitive to acceleration by steric repulsions than it is in nitration, or that in bromination the second step was more sensitive to retardation by steric repulsions than it is in nitration.

Another circumstance which could change the most commonly observed characteristics of the two-stage process of substitution has already been mentioned; it is that in which the step in which the proton is lost is retarded because of a low concentration of base. Such an effect has not been observed in aromatic nitration (§6.2.2), but it is interesting to note that it occurs in *N*-nitration. The *N*-nitration of *N*-methyl-2,4,6-trinitroaniline does not show a deuterium isotope effect in dilute sulphuric acid but does so in more concentrated solutions (> 60% sulphuric acid; k_H/k_D = 4·8).[34]

The cases of pentamethylbenzene and anthracene reacting with nitronium tetrafluoroborate in sulpholan were mentioned above. Each compound forms a stable intermediate very rapidly, and the intermediate then decomposes slowly. It seems that here we have cases where the first stage of the two-step process is very rapid (reaction may even be occurring upon encounter), but the second stages are slow either because of steric factors or because of the feeble basicity of the solvent. The course of the subsequent slow decomposition of the intermediate from pentamethylbenzene is not yet fully understood, but it gives only a poor yield of pentamethylnitrobenzene.[32] The intermediate from anthracene decomposes at a measurable speed to 9-nitroanthracene and the observations are compatible with a two-step mechanism in which $k_{-1} \sim k_2[B]$ and $k_1[NO_2^+] \gg k_{-1}$. There is a kinetic isotope effect (table 6.1), its value for the reaction in acetonitrile being near to the theoretical limit (i.e. $k_2[B]/k_{-1} \ll 1$). The difference between the reaction in acetonitrile and that in sulpholan could arise either because acetonitrile was a weaker base than sulpholan, or because in acetonitrile the transition state leading to the σ-complex was more solvated than in sulpholan. For naphthalene in the same reactions there was no kinetic isotope effect $(k_2[B] \gg k_{-1})$.[33] The contrast between anthracene and naphthalene suggests that steric repulsion may be the dominating feature in the former case.

It is probable that the nitration of anthracene with nitric acid in 7·5 % aqueous sulpholan proceeds through the rapid formation of a complex.[35]

6.4 NITRATION WITH DINITROGEN PENTOXIDE

One mode of substitution occurring when the nitrating system consists of dinitrogen pentoxide in organic solvents involves molecular dinitrogen pentoxide as the effective electrophile (§4.2.3). Evidence that the same electrophile operates when the nitrating system consists of a solution of benzoyl nitrate in carbon tetrachloride has also been given (§5.2).

Substitutions involving dinitrogen pentoxide are not much affected by changes in the polarity of the organic solvent. The series of solvents, carbon tetrachloride, chloroform, acetonitrile and nitromethane produce only a sixfold spread of rate, and this fact is thought to be in accord with the formation of a transition state which is neutral but dipolar, and somewhat more solvated than the neutral reactants. A single-stage bimolecular process was envisaged, possibly with a cyclic transition state (VI).[36] This

(VI)

sort of transition state is made unlikely by the lack of a kinetic isotope effect in the nitration of anisole or toluene with benzoyl nitrate (formed from benzoyl chloride and silver nitrate; table 6.1); it would require opening of the carbon–hydrogen bond in the rate-determining step. The observation does not rule out dinitrogen pentoxide as the electrophile, but shows that loss of hydrogen is kinetically unimportant.[9] Even if some special type of transition state applied to the case of anisole, as has been discussed earlier (§5.3.4), the case is not distinguished from

that of toluene by the study of isotope effects, and it must be admitted
that we have no firm evidence about nitration with dinitrogen pentoxide
so far as the actual substitution is concerned. The general case of substitution by an electrophile $X-Y$, in which X enters the aromatic
molecule and Y^- is eliminated is discussed by Zollinger.[3]

6.5   THE ROLE OF π-COMPLEXES

As well as the σ-complexes discussed above, aromatic molecules combine with such compounds as quinones, polynitro-aromatics and tetra-
cyanoethylene to give more loosely bound structures called charge-
transfer complexes.[37] Closely related to these, but usually known as
π-complexes, are the associations formed by aromatic compounds and
halogens, hydrogen halides, silver ions and other electrophiles.

In π-complexes formed from aromatic compounds and halogens, the
halogen is not bound to any single carbon atom but to the π-electron
structure of the aromatic, though the precise geometry of the complexes
is uncertain.[37a] The complexes with silver ions also do not have the
silver associated with a particular carbon atom of the aromatic ring, as
is shown by the structure of the complex from benzene and silver
perchlorate.[38]

The heats of formation of π-complexes are small; thus, $-\Delta H_{25\,°C}$
for complexes of benzene and mesitylene with iodine in carbon tetrachloride[37b] are 5·5 and 12·0 kJ mol^{-1}, respectively. Although substituent
effects which increase the rates of electrophilic substitutions also increase
the stabilities of the π-complexes, these effects are very much weaker in
the latter circumstances than in the former; the heats of formation just
quoted should be compared with the relative rates of chlorination and
bromination of benzene and mesitylene[39] ($1:3·06 \times 10^7$ and $1:1·89 \times 10^8$,
respectively, in acetic acid at 25 °C).

The solubility of hydrogen chloride in solutions of aromatic hydrocarbons in toluene and in n-heptane at $-78·51$ °C has been measured,
and equilibrium constants for π-complex formation evaluated.[22c] Substituent effects follow the pattern outlined above (table 6.2). In contrast
to σ-complexes, these π-complexes are colourless and non-conducting,
and do not take part in hydrogen exchange.

Given that many electrophiles form π-complexes with aromatic
hydrocarbons, and that such complexes must be present in solutions in
which electrophilic substitutions are occurring, the question arises

TABLE 6.2   *Stabilities of complexes of alkylbenzenes and rates of substitutions*[*]

Substituent	Complexes[†] with:				Chlorination rates[‡]	Nitration rates[§]
	Ag$^+$	ICl	HCl	HF-BF$_3$		
.	0·98	0·36	0·61	.	0·005	0·51
Me	1·04	0·57	0·92	0·01	0·157	0·85
Et	0·86	0·58	1·06	.	0·13	0·82
n-Pr	.	.	.	.	.	0·74
i-Pr	0·89	0·58	1·24	.	0·08	0·67
n-Bu	.	.	.	.	.	0·71
t-Bu	0·79	0·58	1·36	.	0·05	0·60
1,2-Me$_2$	1·26	0·82	1·13	2·0	2·1	0·89
1,3-Me$_2$	1·19	0·92	1·26	20·0	200	0·84
1,4-Me$_2$	1·00	1·00	1·00	1·0	1·00	1·00
1,2,3-Me$_3$	.	.	1·46	~ 40	.	.
1,2,4-Me$_3$	.	.	1·36	40·0	.	.
1,3,5-Me$_3$	0·70	3·04	1·59	2800	80000	1·38

* Relative to p-xylene. The table is based on that given by Olah et al.[11a]
† The first three are π-complexes, the fourth σ-complexes.
‡ From ref. 22c. These values differ slightly from those in ref. 39.
§ For nitration with nitronium tetrafluoroborate in sulpholan.[11a] See table 4.1.

whether π-complexes play any significant role in substitution processes. That they might was first suggested by Dewar,[40a] who, however, now shares the view that in the majority of substitutions the rate-determining step leads to the arenonium ion, the σ-complex already discussed.[40b] However, he criticises the description of complexes, such as those mentioned already in this section, as π-complexes, maintaining that in a true π-complex the π-donor and acceptor are linked by a covalent bond.[40b]

We have seen (§6.2.3) that there is a close relationship between the rates of electrophilic substitutions and the stabilities of σ-complexes, and facts already quoted above suggest that no such relationship exists between those rates and the stabilities of the π-complexes of the kind discussed here. These two contrasting situations are further illustrated by the data given in table 6.2. As noted earlier, the parallelism of rate data for substitutions with stability data for σ-complexes is not limited to chlorination (§6.2.4). Clearly, π-complexes have no general mechanistic or kinetic significance in electrophilic substitutions.

The work of Olah et al. on nitration with nitronium salts in organic solvents has already been discussed in some of its aspects (§4.4). It will be recalled that it attracted interest for two reasons; the almost complete disappearance of intermolecular selectivities, and the retention of intramolecular selectivities (tables 4.1, 4.2, 6.2). The combination of these circumstances produced unusual partial rate factors (table 4.2). On the basis of these results Olah postulated that in nitrations carried out by this method the normal mechanism of substitution was replaced by one in which the rate-determining step was the formation of a π-complex. We believe that the available evidence is now sufficient to demonstrate conclusively that the disappearance of intermolecular selectivities was a consequence of slow mixing (§4.4.2, 4.4.3); if the intramolecular selectivities were determined by the normal mechanism of substitution the derivation of partial rate factors is invalidated, and the values obtained require no explanation.

It has been observed in connection with nitrations of reactive compounds occurring at the encounter rate in sulphuric acid (§2.5)[41] that 'in a situation where selectivity between different substrates has disappeared, a high degree of positional selectivity is still maintained (e.g. naphthalene, toluene, phenol, and o-xylene)...This makes it difficult to imagine that the first interaction between the nitronium ion and the aromatic ring occurs at one or other particular nuclear position.' Such a view would almost necessarily lead to the idea that the 'encounter-pair' through which reaction is envisaged to be occurring was a structure of the same kind as the π-complexes just discussed. The rate-determining formation of the encounter pair would then be followed by product-forming steps giving rise to isomers in proportions determined by individual positional reactivities.

If, on the other hand, the encounter pair were an oriented structure, positional selectivity could be retained for a different reason and in a different quantitative sense. Thus, a monosubstituted benzene derivative in which the substituent was sufficiently powerfully activating would react with the electrophile to give three different encounter pairs; two of these would more readily proceed to the substitution products than to the starting materials, whilst the third might more readily break up than go to products. In the limit the first two would be giving substitution at the encounter rate and, in the absence of steric effects, products in the statistical ratio whilst the third would not. If we consider particular cases, there is nothing in the rather inadequate data available to discourage the view that, for example, in the cases of toluene or phenol, which in sulphuric acid are nitrated at or near the encounter rate,[41] the *meta*-positions are distinguished by not being subject to encounter control. In that case, they would differ from the *ortho*- and *para*-positions, which would themselves react in the statistical ratios of 2:1.

A careful study of isomer distributions might thus provide information about the structure of the encounter pair.

REFERENCES

1. Fieser, L. F. (1943). In *Organic Chemistry: An Advanced Treatise*, 2nd ed., vol. 1, ed. H. Gilman. New York: Wiley.
2. Berliner, E. (1964). In *Progress in Physical Organic Chemistry*, vol. 2, ed. S. G. Cohen, A. Streitwieser and R. W. Taft. New York: Interscience.
3. Zollinger, H. (1964). In *Advances in Physical Organic Chemistry*, vol. 2, ed. V. Gold. London: Academic Press.
4. Hughes, E. D., Ingold, C. K. & Reed, R. I. (1950). *J. chem. Soc.* p. 2400.
5. Bennett, G. M., Brand, J. C. D., James, D. M., Saunders, T. G. & Williams, G. (1947). *J. chem. Soc.* p. 474.
6. Melander, L. (a) (1960). *Isotope Effects on Reaction Rates*. New York: Ronald Press. (b) (1950). *Ark. Kemi* 2, 211.
7. Hammett, L. P. (1940). *Physical Organic Chemistry*. New York: McGraw-Hill.
8. Hammond, G. S. (1955). *J. Am. chem. Soc.* 77, 334.
9. Halvarson, K. & Melander, L. (1957). *Ark. Kemi* 11, 77.
10. Lauer, W. M. & Noland, W. E. (1953). *J. Am. chem. Soc.* 75, 3689.
11. Olah, G. A., Kuhn, S. J. & Flood, S. H. (a) (1961). *J. Am. chem. Soc.* 83, 4571; (b) p. 4581.
12. Myhre, P. C. (a) (1960). *Acta chem. scand.* 14, 219; (b) with Beug, M. & James, L. L. (1968). *J. Am. chem. Soc.* 90, 2105.
13. Gleghorn, J., Moodie, R. B., Schofield, K. & Williamson, M. J. (1966). *J. chem. Soc.* B, p. 870.
14. Bonner, T. G., Bowyer, F. & Williams, G. (1953). *J. chem. Soc.* p. 2650.
15. Oestman, B. (1962). *Ark. Kemi* 19, 499.
16. Eastham, J. F., Bloomer, J. L. & Hudson, F. M. (1962). *Tetrahedron* 18, 653.
17. Suhr, H. & Zollinger, H. (1961). *Helv. chim. Acta* 44, 1011.
18. Swain, C. G., Knee, T. E. C. & Kresge, A. J. (1957). *J. Am. chem. Soc.* 79, 505.
19. Cowdrey, W. A., Hughes, E. D., Ingold, C. K., Masterman, S. & Scott, A. D. (1937). *J. chem. Soc.* p. 1252.
    Hughes, E. D. & Ingold, C. K. (1941). *J. chem. Soc.* p. 608.
20. de la Mare, P. B. D. & Ridd, J. H. (1959). *Aromatic Substitution; Nitration and Halogenation*, p. 72. London: Butterworths.
21. Ingold, C. K. (1953). *Structure and Mechanism in Organic Chemistry*, p. 281. London: Bell.
22. Brown, H. C. (a) with Pearsall, H. W. (1952). *J. Am. chem. Soc.* 74, 191; (b) with Wallace, W. J. (1953). *J. Am. chem. Soc.* 75, 6268; (c) with Brady, J. D. (1952). *J. Am. chem. Soc.* 74, 3570.

23. Norris, J. F. & Ingraham, J. N. (1940). *J. Am. chem. Soc.* **62**, 1298.
    Baddeley, G., Holt, G. & Voss, D. (1952). *J. chem. Soc.* p. 100.
    Eley, D. D. & King, P. J. (1952). *J. chem. Soc.* pp. 2517, 4972.
24. (a) McCaulay, D. A. & Lien, A. P. (1951). *J. Am. chem. Soc.* **73**, 2013.
    (b) Kilpatrick, M. & Luborsky, F. E. (1953). *J. Am. chem. Soc.* **75**, 577.
    (c) Olah, G. & Kuhn, I. (1958). *J. Am. chem. Soc.* **80**, 6535.
25. MacLean, C., van der Waals, J. H. & Mackor, E. L. (1958). *Molec. Phys.*
    1, 247.
26. Verrijn Stuart, A. A. & Mackor, E. L. (1957). *J. chem. Phys.* **27**, 826.
    Dallinga, G., Mackor, E. L. & Verrijn Stuart, A. A. (1958). *Molec. Phys.* **1**,
    123.
27. Gold, V. & Tye, F. L. (1952). *J. chem. Soc.* p. 2172.
28. Brown, H. C. & Nelson, K. L. (1955). In *The Chemistry of Petroleum,
    Hydrocarbons*, vol. 3, chap. 56, ed. B. T. Brooks, S. S. Kurtz, C. E. Boord
    and L. Schmerling. New York: Reinhold.
29. Condon, F. E. (1952). *J. Am. chem. Soc.* **74**, 2528.
30. Streitwieser, A. (1961). *Molecular Orbital Theory for Organic Chemists*,
    New York: Wiley. (a) p. 318, (b) p. 326–7.
31. Olah, G. A. & Kuhn, S. J. (1958). *J. Am. chem. Soc.* **80**, 6541.
32. (a) Kreienbühl, P. & Zollinger, H. (1965). *Tetrahedron Lett.* p. 1739.
    (b) Hanna, S. B., Hunziker, E., Saito, T. & Zollinger, H. (1969). *Helv.
    chim. Acta* **52**, 1537.
33. Cerfontain, H. & Telder, A. (1967). *Recl Trav. chim. Pays-Bas Belg.* **86**,
    370.
34. Halevi, E. A., Ron, A. & Speiser, S. (1965). *J. chem. Soc.* p. 2560.
35. Hoggett, J. G., Moodie, R. B. & Schofield, K. (1969) *J chem. Soc.* B, p. 1.
36. Gold, V., Hughes, E. D., Ingold, C. K. & Williams, G. H. (1950). *J. chem.
    Soc.* p. 2452.
37. Foster, R. (1969). *Organic Charge-Transfer Complexes.* New York: Academic
    Press. (a) p. 230; (b) p. 205.
38. Rundle, R. E. & Goring, J. H. (1950). *J. Am. chem. Soc.* **72**, 5337.
    Smith, H. G. & Rundle, R. E. (1958). **80**, 5075.
39. Baciocchi, E. & Illuminati, G. (1967). *Progress in Physical Organic Chemistry*,
    (ed. A. Streitwieser and R. W. Taft), vol. 5. New York: Interscience.
40. Dewar, M. J. S. (a) (1949). *The Electronic Theory of Organic Chemistry.*
    Oxford University Press. (b) with Marchand, A. P. (1965). *Ann. Rev. phys.
    Chem.* **16**, 321.
41. Coombes, R. G., Moodie, R. B. & Schofield, K. (1968). *J. chem. Soc.* B,
    p. 800.

7 Nitration and aromatic reactivity: A. The theoretical background

7.1 INTRODUCTION

It is the purpose of this chapter to provide a résumé of theoretical concepts which are used in discussing aromatic reactivity. Extended discussion is unnecessary, for many of the concepts are the common currency of organic chemistry, and have been frequently expounded.[1-8]

7.1.1 The measurement of aromatic reactivity

Most of the reactions with which organic chemists are concerned involve poly-atomic molecules, and occur in solution at temperatures not far removed from the ambient. There is not at present the faintest possibility of chemical theory predicting the absolute rates of such processes.

There are available from experiment, for such reactions, measurements of rates and the familiar Arrhenius parameters and, much more rarely, the temperature coefficients of the latter. The theories which we use, to relate structure to the ability to take part in reactions, provide static models of reactants or transition states which quite neglect thermal energy. Enthalpies of activation at zero temperature would evidently be the quantities in terms of which to discuss these descriptions, but they are unknown and we must enquire which of the experimentally available quantities is most appropriately used for this purpose.

In transition state theory, the rate constant, k, is given by the following expression:

$$k = \frac{\mathbf{k}T}{h} \exp\left(-\Delta G^{\ddagger}/RT\right).$$

Admitting the impossibility of calculating absolute rates, we can concern ourselves with the effect of a structural modification to a particular reactant which we take as a point of reference; if the rate constant for the reaction involving the modified compound is k, and that for the reference substance k_0, then:

$$2 \cdot 303\, RT \log_{10} k/k_0 = \Delta G_0^{\ddagger} - \Delta G^{\ddagger}.$$

Arguments have been presented[9] that this difference in changes in Gibbs' function, rather than the similar difference in enthalpies of activation, $\Delta H_0^{\ddagger} - \Delta H^{\ddagger}$, better represents the quantity with which theoretical treatments of structural effects on reactivity are concerned, namely differences in the zero energy* of activation. Rate constants of reaction are, then, the data to use in testing theoretical predictions, but small changes in rate must be treated cautiously.

The behaviour of benzene is the datum from which any discussion of aromatic compounds must start: *the reactivity of an aromatic compound is its rate of reaction relative to that of benzene when both are taking part in reactions occurring under the same conditions and proceeding by the same mechanism.*

7.1.2 *Limits to the meaning of aromatic reactivity*

The above definition implies that the reactivity of an aromatic compound depends upon the reaction which is used to measure it, for the rate of reaction of an aromatic compound relative to that for benzene varies from reaction to reaction (table 7.1). However, whilst a compound's reactivity can be given no unique value, different substitution reactions do generally set aromatic compounds in the same sequence of relative reactivities.

As a means of studying the reactivities of aromatic compounds towards electrophiles, nitration has one major advantage compared with other substitution processes; over a wide range of experimental conditions it involves the same electrophile, the nitronium ion, and so is applicable to a large range of compounds, differing widely not only in reactivity but in such practically important properties as solubility. In very varied conditions of nitration an aromatic compound shows surprisingly similar reactivities (see the data for toluene; table 4.1, columns *b–g, k–o, q*). Nitration in aqueous sulphuric acid can provide data for compounds covering a very large span of reactivities for, as has been seen (§2.4.2), the second-order rate coefficient decreases by a factor of about 10^4 for each decrease of 10 % in the concentration of the sulphuric acid.

Generally the determination of the reactivity of a particular compound depends upon comparison of its rate of nitration with that of benzene at the same acidity and temperature. Because of the spread of rates this may not be practically possible and, in any case, is usually not necessary because of the parallelism existing among rate profiles (fig. 2.4). Reactivities in aqueous sulphuric acid are, in fact, very nearly independent of acidity, and stepwise comparison of data for a compound with those of benzene determined at different acidities is possible.

* Zero energy is the energy which a species would have at absolute zero in the absence of zero-point vibrational energy.

TABLE 7.1 *Partial rate factors for some electrophilic*
substitutions of toluene

Reaction	Reagent	f_o	f_m	f_p
Nitration*	$HNO_3/AcOH/25\ °C$	49	2·4	70
	$HNO_3/MeNO_2/25\ °C$	49	2·5	56
	$HNO_3/Ac_2O/0\ °C$	39	3·0	51
Chlorination[13]	$Cl_2/aq.\ AcOH$	534	.	552
	$Cl_2/AcOH$	617	5	820
	$Cl_2/MeCN$	1830	9·1	6250
	$HClO/H_2O/H^+$ (i.e. Cl^+)	134	4·0	82
Bromination[13]	$Br_2/aq.\ AcOH$	600	5·5	2420
	$HBrO/H_2O/H^+$ (i.e. Br^+)	76	25	59

* See table 4.2.

There are certain limitations to the usefulness of nitration in aqueous sulphuric acid. Because of the behaviour of the rate profile for benzene, comparisons should strictly be made below 68 % sulphuric acid (§2.5; fig. 2.5); rates relative to benzene vary in the range 68–80 % sulphuric acid, and at the higher end of this range are not entirely measures of relative reactivity.[10a] For deactivated compounds this limitation is not very important,[11] but for activated compounds it is linked with a fundamental limit to the significance of the concept of aromatic reactivity; as already discussed (§2.5), nitration in sulphuric acid cannot differentiate amongst compounds not less than about 38 times more reactive than benzene. At this point differentiation disappears because reactions occur at the encounter rate.

For deactivated compounds this limitation does not exist, and nitration in sulphuric acid is an excellent method for comparing the reactivities of such compounds. For these, however, there remains the practical difficulty of following slow reactions and the possibility that with such reactions secondary processes might become important. With deactivated compounds, comparisons of reactivities can be made using nitration in concentrated sulphuric acid; such comparisons are not accurate because of the behaviour of rate profiles at high acidities (§2.3.2; figs. 2.1, 2.3).

The limit to the significance of aromatic reactivity set by reaction at the encounter rate is reached at different levels of reactivity in different conditions. As already seen, for nitration with nitric acid in organic solvents the limit appears at reactivities 300–400 times that of benzene (§3.3). This fact extends slightly the range of activated compounds which can usefully be studied by nitration, but nitration in organic solvents is less useful than nitration in sulphuric acid for deactivated compounds, being slower.

Reaction at the encounter rate sets a limit to the meaning of reactivity no matter what reaction is used to measure the latter. Surprisingly, the only other electrophilic substitution, besides nitration, in which this limit has been identified as having been approached or reached is that of molecular bromination in aqueous mineral acid.[12] Some anilines clearly react at or near the encounter rate and, although the point has not been fully examined, it is clear that for bromination the level of differentiable reactivity is much higher than in nitration.[13] In this respect, bromination is a more suitable reaction than is nitration for studying the reactivities of activated compounds, but of course it has its peculiar disadvantages: variation of effective electrophile with conditions; complicated kinetics and solubility problems in some conditions; dependence of which step is rate-determining upon conditions; slowness of reaction with deactivated compounds, and so on.

The significance of establishing a limiting rate of reaction upon encounter for mechanistic studies has been pointed out (§2.5). In studies of reactivity, as well as setting an absolute limit to the significance of reactivity in particular circumstances, the experimental observation of the limit has another dependent importance: *if further structural modification of the aromatic compound leads ultimately to the onset of reaction at a rate exceeding the observed encounter rate then a new electrophile must have become operative, and reactivities established 'above the encounter rate' cannot properly be compared with those measured below it.*

7.2   THEORIES RELATING STRUCTURE AND REACTIVITY

7.2.1   *The electronic theory of organic chemistry*

This qualitative theory still provides the most widely used means for describing reactions in organic chemistry.[1,2] Two principal modes of electronic interaction in organic molecules are recognised; the inductive and mesomeric effects.

Three aspects of the inductive effect have to be considered:[1,3,7b,14a] the σ-inductive effect, the inducto-electromeric or π-inductive effect, and the direct field effect. The first of these is the one most frequently used. It arises from the unsymmetrical distribution of the electrons of the σ-bonds in molecules possessing polar or dipolar groups. The effect is transmitted within the molecule by the successive distortion of the σ-bonds; atoms or groups more electronegative than carbon attract electrons to themselves at the expense of the carbon atoms to which they are attached, whereas atoms or groups more electropositive than carbon tend to lose electrons to the rest of the molecule. However, for convenience, the electronic character of a group is referred to that of hydrogen in a similar molecular situation; groups which attract electrons more than does hydrogen are considered to exert a $-$ I effect, and groups which repel electrons more than does hydrogen exert a $+$ I effect. There is much evidence to show that the σ-inductive effect decays very rapidly with distance from the substituent.[3, 7b, 15]

The π-inductive effect describes how an inductive substituent might selectively influence the electron distribution at the o- and p-positions of the aromatic nucleus. A familiar example is represented by the

(I)　　　　　　(II)　　　　　　　　(III)　　　·　　(IV; $X = \overset{+}{N}H_3$ or CO_2H)

symbol (I) for the case of the methyl group, and another by (II) showing how the nitrogen atom of a pyridine nucleus can stabilise the anion of β-picoline.[16]

Excluding the phenomenon of hyperconjugation, the only other means by which electronic effects can be transmitted within saturated molecules, or exerted by inductive substituents in aromatic molecules, is by direct electrostatic interaction, the direct field effect. In early discussions of substitution this was usually neglected for qualitative purposes since it would operate in the same direction (though it would be expected to diminish in the order *ortho* > *meta* > *para*) as the σ-inductive effect and assessment of the relative importance of each is difficult: however, the field effect was recognised as having quantitative significance.[1]

In contrast, equilibrium properties have been successfully discussed in terms of the field effect. Notable instances are those of the ionisation constants of saturated dibasic acids,[1, 3] and of carboxyl groups held in rigid, saturated rings (III).[17a] The importance of the effect for the ionisation of groups in charged and dipolar aromatic molecules has also been urged.[14b, 17b] Measurements of the ionisation constants of the ions (IV) are said to provide no support for the selective relay of the inductive effect to the *o*- and *p*-positions[17b] as visualised with the π-inductive effect and outlined above. The results were held to show that the inductive effect was best regarded as falling off smoothly with distance, possibly in accord with the Coulomb law. On the basis of these results the orientation of electrophilic substitution into phenyltrimethylammonium ion was thought to be best described in terms of electrostatic repulsions present in the transition states (see below). The interpretation of the ionisation constants of (IV) leading to these conclusions is disputed.[1]

In unsaturated molecules electronic effects can be transmitted by mesomerism as well as by inductive effects. As with the latter, the mesomeric properties of a group are described by reference to hydrogen. Groups which release electrons to the unsaturated residue of the molecule are said to exert a $+M$ effect, whereas groups which attract electrons are said to exert a $-M$ effect. In aromatic structures the important feature of an M-substituent is that it influences the *o*- and *p*-positions selectively.

The electronic theory provides by these means a description of the influence of substituents upon the distribution of electrons in the ground state of an aromatic molecule as it changes the situation in benzene. It then assumes that an electrophile will react preferentially at positions which are relatively enriched with electrons, providing in this way an 'isolated molecule' theory of reactivity.

However, the electronic theory also lays stress upon substitution being a developing process,[1, 18, 19] and by adding to its description of the polarization of aromatic molecules means for describing their polarisability by an approaching reagent, it moves towards a transition state theory of reactivity. These means are the electromeric and inductomeric effects.

The electromeric effect (E) of a group, the time-dependent distortion of its mesomeric effect, is like the mesomeric effect in being able to affect only alternate atoms from the substituted atom. Groups showing strong $+M$ tendencies can exert strong $+E$ effects if needed by the reaction, but will only yield a weak $-E$ effect should the electronic requirements of the reaction be of the opposite sign. Similarly, groups possessing strong $-M$ character will give $-E$ assistance strongly, or $+E$ assistance weakly, according to the requirements of the reaction. The combined mesomeric and electromeric effects are termed conjugative or resonance effects, descriptions now preferable to that (tauto-meric) originally used by Ingold.

There is little evidence for the operation in reactions of the inducto-meric effect, the time-dependent analogue of the inductive effect. This may be so because the electrons of the σ-bonds are more localized and more tightly bound than the electrons of the delocalized system, and are thus not so susceptible to the demands of the reagent.

A familiar feature of the electronic theory is the classification of substituents, in terms of the inductive and conjugative or resonance effects, which it provides. Examples from substituents discussed in this book are given in table 7.2. The effects upon orientation and reactivity indicated are only the dominant ones, and one of our tasks is to examine in closer detail how descriptions of substituent effects of this kind meet the facts of nitration. In general, such descriptions find wide acceptance, the more so since they are now known to corres-pond to parallel descriptions in terms of molecular orbital theory (§§7.2.2, 7.2.3). Only in respect of the interpretation to be placed upon the inductive effect is there still serious disagreement. It will be seen that recent results of nitration studies have produced evidence on this point (§9.1.1).

TABLE 7.2   *The classification of substituents**

Type	Electronic mechanisms	Effect on:		Example
		Orientation	Reactivity	
$+I$	Ph $\leftarrow$ Y	$o:p$	Activation	Ph.Me
$-I$	Ph $\rightarrow$ Y	m	Deactivation	$\overset{+}{\text{Ph.NH}_3}$
$-I-R$	Ph $\overset{\frown}{\rightarrow}$ Y	m	Deactivation	$\begin{cases} \text{Ph.NO}_2 \\ \text{Ph.CO}_2\text{Et} \end{cases}$
$-I+R$	Ph $\overset{\frown}{\rightarrow}$ Y	$\begin{cases} m \\ o:p \\ o:p \end{cases}$	Deactivation Deactivation Activation	$\overset{+}{\text{Ph.SMe}_2}$ Ph.Cl Ph.OMe
$+I+R$	Ph $\overset{\frown}{\leftarrow}$ Y	$o:p$	Activation	$\overset{-}{\text{Ph.O}}$

* Based on a table given by Ingold.[1] The symbol R (resonance effect)[3] is used in place of T (see text).

In providing an isolated molecule description of reactivity, qualitative resonance theory[6] is roughly equivalent to that given above, but is less flexible in neglecting the inductive effect and polarisability.[19] It is most commonly used now as a qualitative transition state theory, taking the

Wheland intermediate (see below) as its model for the transition state. In this form it is illustrated by the case mentioned above, that of nitration of the phenyltrimethylammonium ion.[17b] For this case the transition state for *m*-nitration is represented by (v) and that for *p*-substitution by (vi). It is argued that electrostatic repulsions in the former are smaller than in the latter, so that *m*-nitration is favoured, though it is associated with deactivation. Similar descriptions can be given for the gross effects of other substituents upon orientation.

(V)

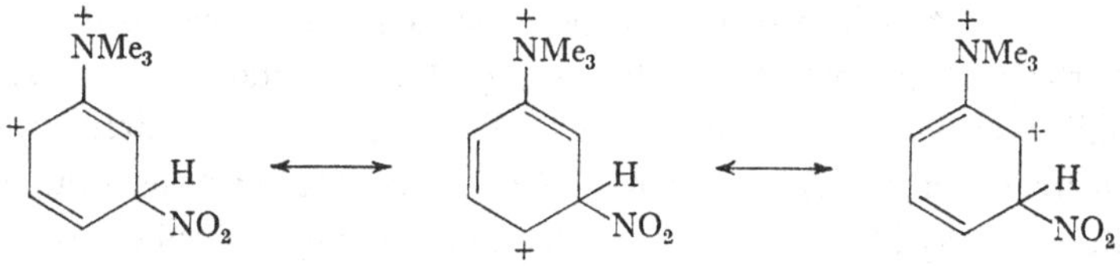

(VI)

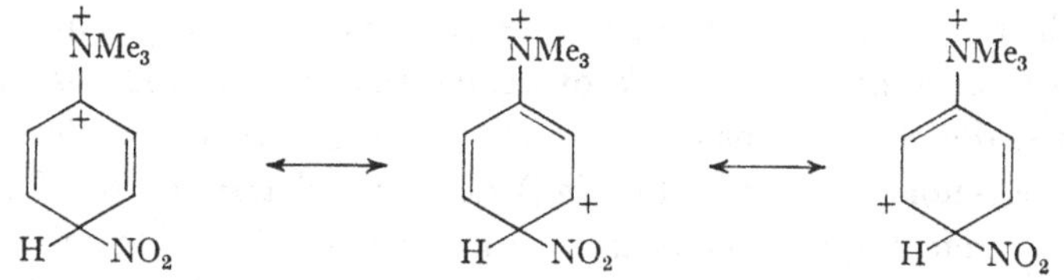

It is still the case that most M.O. (molecular orbital) treatments of aromatic reactivity use the Hückel approximation.[3-8] The energy difference which, within the limits already discussed, measures aromatic reactivity, can be divided into three parts: one relating to change in π-electronic energy, another to changes in the energies of the σ-bonded structures, and another to electrostatic effects. Changes in σ-bonds are neglected since they are likely to be similar for similar compounds undergoing the same reaction. The problem of calculating the reactivity of an aromatic compound is thus reduced to that of calculating for it the change between ground and transition states of π-electron and electrostatic energies, as compared with that for benzene. As implied above, two extreme models of the transition state can be considered; in the isolated molecule treatment the transition state is taken to resemble the ground states of the reactants, whilst at the other extreme the Wheland intermediate (see below) is used to represent the transition state.

Aromatic reactivity: A. Theoretical background

7.2.2 M.O. theory and the isolated molecule treatment

In this model, reaction is considered to occur preferentially at that position in the aromatic molecule to which the approach of the electrophile causes the smallest increase in zero energy. In molecules possessing polar or dipolar groups, long range electrostatic forces will initially be the most important.

The π-electrons are not distributed uniformly amongst the various positions in non-alternant hydrocarbons or heteromolecules;* in their ground states the various positions are associated with different values of the π-electronic charge density, q_r. The qualitative idea that electrophiles undergo reaction more easily at positions which are rich in electrons, suggests a correlation[20] between rates and the value of q_r. In alternant hydrocarbons all the positions of the molecule have a charge density of unity so that consideration of the ground state indicates that all the positions should be equally reactive. However, experimentally this is not generally the case. To overcome this difficulty it was suggested that if the charge densities were the same, differences in the values of the self-atom polarizabilities (π_{rr}) determined the course of reaction. This parameter, which reflects the ease with which the incoming electrophile changes the electron density at a position, is discussed below. In this form the theory obviously parallels the electronic theory already described with the inclusion of considerations of polarisability.

These parameters, q_r and π_{rr}, are two of a number of such parameters whose values are used as indices of reactivity in electrophilic aromatic substitution.[3-5,7] However, they are not completely independent quantities as the following discussion shows.

In the Hückel theory, the π-electron energy of a conjugated molecule can be expressed by the following equation:

$$E_\pi = \sum_r q_r \alpha_r + 2 \sum_r \sum_t p_{rt} \beta_{rt}$$
$$\text{(all atoms bonded to } r\text{)}$$

The change in E_π induced by the approaching electrophile can be simulated by changing the α and β terms of the atom undergoing reaction. The following equation represents such changes.

$$\delta E_\pi = q_r \delta \alpha_r + \tfrac{1}{2}\pi_{rr}(\delta \alpha_r)^2 + \ldots + 2p_{rt}\delta\beta_{rt} + \ldots + \text{other terms.}$$

---

* This term denotes all compounds with substituents containing atoms other than carbon, and includes heterocyclic compounds.

Physically a change in α is caused by the electrostatic field of the reagent, and a change in β by the modification of the atom from an sp^2 to an sp^3 state, with the concomitant isolation of the atom from the delocalized system. In considering the approach of electrophilic reagents, which usually have marked polar properties, considerable changes in α occur long before any significant changes in hybridisation.

As pointed out above, aromatic reactivity depends, at least in part, on the way in which the π-electron energies of the molecules change between the ground state and the transition state. The last equation gives a measure of this change, over the early part of the reaction where the molecule is not too seriously distorted from its ground state. As the electrophile approaches the site of reaction, q_r reflects to a first approximation the change in the π-energy of the aromatic, and because $\delta\alpha_r$ is negative, reaction is favoured by high values of q_r. On the closer approach of the reagent the term involving π_{rr} assumes importance. The effect on δE_π of these terms together can only be assessed by using arbitrary values of $\delta\alpha_r$ in the last equation. Higher polarizability terms become more important as the electrophile approaches even more closely but their calculation is lengthy.

The use of q_r and π_{rr} separately as reactivity indices can lead to misleading results. Thus, whilst within the approximations used, the use of either separately leads to the same conclusions regarding electrophilic substitution into halogenobenzenes (§9.1.4), the orientation of substitution in quinoline (§9.4.2) cannot be explained even qualitatively using either alone. By taking the two in combination, it can be shown that as the values of $\delta\alpha_r$ are progressively increased to simulate reaction, the differences in δE_π explain satisfactorily the observed orientation.[21]

7.2.3 *M.O. theory and the transition state treatment*

In 1942 Wheland[22] proposed a simple model for the transition state of electrophilic substitution in which a pair of electrons is localised at the site of substitution, and the carbon atom at that site has changed from the sp^2 to the sp^3 state of hybridisation. Such a structure, originally proposed as a model for the transition state is now known to describe the σ-complexes which are intermediates in electrophilic substitutions (§6.2.3).

As we have seen, the important zero energy difference which measures aromatic reactivity contains a term involving π-electron energies, and with the transition state model there will also be a contribution from the formation of the bond between the reagent and the aromatic compound. For substitution in related systems the latter term is neglected and interest centred on the difference between the π-electron energy of the starting material and the Wheland model of the transition state, as compared with the same difference for benzene. These differences are called localisation energies, that for atom r in an electrophilic substitution being denoted L_r^+.

Localisation energies have been calculated for a number of cases, most commonly in the Hückel approximation; even for large molecules they can now be readily calculated using digital computers. However, a method due to Dewar[7, 23a] permits the calculation of approximate localisation energies, called reactivity numbers, by simple arithmetic from the coefficients of the non-bonding M.O.'s of the atoms adjacent to the site of attack. These reactivity numbers (N_r), although numerically smaller than Hückel localisation energies, show a good linear correlation with them.[4]

Reactivity numbers of the most reactive positions have been used to correlate the reactivities in nitration (see below)[23b] and other substitutions[4] of a series of polycyclic aromatic hydrocarbons, and they give somewhat better correlations than any of the other commonly used indices of reactivity.[4a] The relationship shown below, which was discussed earlier (§7.1.1),

$$2 \cdot 303 RT \log_{10} k/k_0 = \Delta G_0^{\ddagger} - \Delta G^{\ddagger}$$

can be rewritten in terms of partial rate factors (f_r) and reactivity numbers (or localisation energies) as follows, where f_r and N_r refer to

$$2 \cdot 303 RT \log_{10} f_r = -\beta(N_r - N_B)$$

positions in benzene derivatives or in polynuclear compounds, N_B is the reactivity number for benzene, and β is the Hückel resonance integral. Plots of $\log_{10} f_r$ $v.$ ($N_r - N_B$) for the results of nitrating and chlorinating polynuclear hydrocarbons were approximately linear. However, the slopes of these two correlations differed, and both slopes were less than might have been expected from the usual value of β estimated from comparisons of empirical and calculated resonance energies in aromatic molecules.[13] Two qualifications must be made about this latter circumstance. First that the Hückel theory is an approximation, and that no great significance should be given to the numerical values of the parameters used in it, or the quantitative predictions made by it.

Secondly, the use of a value of the resonance integral β derived from empirical resonance energies in other contexts is not justifiable.

The fact that the ratios of rates were much greater in chlorination than in nitration, prompted Dewar to suggest that the actual transition state was intermediate between the Wheland model and the isolated molecule model.[23b] He accommodated this variation in the relative rates within his discussion by treating β as a variable whose value depended on the nature of the reaction. With the notation that β_R ($< \beta$) is the empirical parameter dependent on the reaction, the equation for the partial rate factor takes the following form:

$$2\cdot303RT \log_{10} f_r = -\beta_R(N_r - N_B).$$

His data suggested values for β_R of -12 and -6 kcal mol^{-1} for molecular chlorination and nitration respectively, indicating that the transition states in nitration resemble the reactants more than do the transition states in chlorination.

The general method of using a correlation of reaction rates with localisation energies or reaction numbers to derive a value of β_R is regarded as a means of studying the structures of transition states.[7a] Empirically, the value of the parameter obtained from the correlation of localisation energies with equilibrium constants for the protonation of aromatic hydrocarbons in anhydrous hydrogen fluoride sets a rough limit which should be reached in a substitution reaction with a transition state close to the Wheland intermediate. Interesting observations on the method, arising from studies of halogenation, have been made by Mason.[24]

Dewar's treatment of transition state structure, using reactivity numbers, has the logical defect that in the intermediate kinds of transition states for which it provides evidence the electron localisation is only partial. However, in obtaining the values of the reactivity numbers (which are approximate localization energies), the process of localization is considered to be complete; thus, values of parameters which strictly are relevant only to the Wheland type of transition state are incorporated into a different model.[25]

Most correlations of rates with localisation energies have used values for the latter derived from the Hückel approximation. More advanced methods of M.O. theory can, of course, be used, and fig. 7.1 illustrates plots correlating data for the nitration of polynuclear hydrocarbons in acetic anhydride[23b,c] with localisation energies derived from self-

Aromatic reactivity: A. Theoretical background

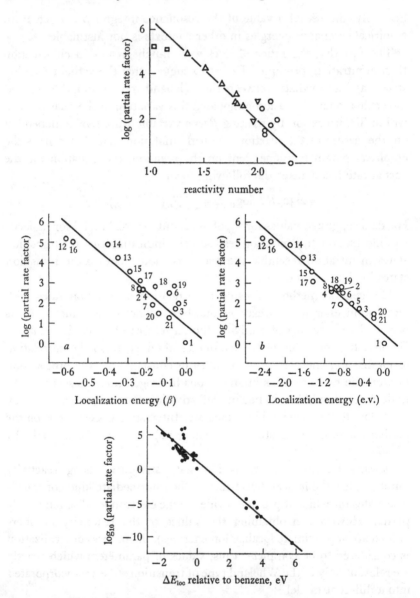

Fig. 7.1. Nitration of polynuclear hydrocarbons by nitric acid in acetic anhydride. (i) Plot of $\log_{10}(K_i/K_0)$ against N. $\square$, 9-Anthryl positions; $\triangle$, α-napthyl positions; ∇, 4-phenanthryl positions; $\bigcirc$, other positions. (From Dewar et al.[23b]) (ii) and (iii) Relative electrophilic localization energies vs. logarithms of partial rate factors for nitration: (a) Hückel, (b) PPP with fixed β. (From Dewar & Thompson.[23c]) (iv) Plot of $\log K$ vs. ΔE_{loc}. (From Dewar.[23c])

consistent field M.O. theory and from Hückel theory, and with Dewar's reactivity numbers. The nitration of polynuclear hydrocarbons in acetic anhydride has already been discussed (§§5.3.2, 5.3.3) and it is most probable that the results for the most reactive ones studied are vitiated by the occurrence of nitrosation. This is also very probably the case with the very reactive phenols and amines studied (table 5.3). For these reasons the correlation of data for the hydrocarbons with reactivity numbers, and the extended correlation including ethers and amines are of doubtful significance.[10]

As indicated, the indices of reactivity mentioned in this section (q_r, π_{rr}, and $L_r{}^+$) are only three from among several that have been used. These and their interrelationships have been critically discussed.[21] It is a common feature of M.O. theories, whatever their degree of sophistication, that they can deal more easily with hydrocarbons than with heteromolecules. In treating the latter the influence of the hetero-atoms has to be dealt with by ascribing numerical values to adjustable parameters, and sometimes, in addition to this, by ascribing numerical values to the so-called auxiliary inductive parameters in an attempt to simulate the σ-inductive effect.[4] The problem has been much discussed, especially in connection with Hückel theory.[4] Because of these difficulties it is a pity, from the point of view of seeking correlations between experiment and M.O. theory, that so much more effort has been given to studying substituent effects in benzene derivatives than in comparing benzene with its polynuclear analogues.

It is generally the case, and will be later illustrated by examples, that the treatment of aromatic reactivity through localisation energies performs well in predicting orientation of substitution, but much less well in dealing with state of activation. It may be that this deficiency will be removed by the use of M.O. theories more adequate than the Hückel approximation.[21] Dewar's reactivity numbers and the use of different values of β_R to characterise different reactions is an attempt to cope with the fact already stressed (§7.1.2), that the reactivity of an aromatic compound depends on the reagent with which it is interacting. This aspect is now beginning to attract theoretical discussion.[26]

7.2.4 *Electrostatic theories*

The isolated molecule treatment of reactivity, which, in both the electronic theory and in M.O. theory, attempts to predict the site of electrophilic substitution from a consideration of the electron densities at the possible positions, is essentially an electrostatic treatment. If the electrophile is a cation, this type of theory is really concerned with charge interactions; if it is a neutral molecule, with pole–dipole interactions.

Attempts to incorporate electrostatic effects into theories of electrophilic substitution in a more quantitative way have been made from time to time. In the earliest of these Ri and Eyring[27] proceeded by assuming that the change in rate of nitration of a mono-substituted derivative of benzene, compared with that of benzene, arises from a change in the Gibbs' function of activation due to the energy of electrostatic interaction in the transition state between the charge of the nitronium ion (e_n) and that of the carbon atom being attacked (e_y). They consequently wrote

$$k_y = \frac{kT}{h} \exp\left[-\left(\Delta G^{\ddagger} + \frac{e_y e_n}{rD}\right) \Big/ RT\right],$$

and thence by reference to benzene $\log(k_y/k_0) = -e_y e_n/2 \cdot 303 r DRT$. Thus, from rate data for nitration they were able to calculate charge distributions in benzene derivatives, and from these, dipole moments which to compare with experimental values. They were also able to calculate isomer proportions from dipole moments. Subsequently, Kenner[28] applied a form of this treatment to a study of the properties of the halogenobenzenes.

The model adopted by Ri and Eyring is not now acceptable, but some of the more recent treatments of electrostatic effects are quite close to their method in principle. In dealing with polar substituents some authors have concentrated on the interaction of the substituent with the electrophile whilst others have considered the interaction of the substituent with the charge on the ring in the transition state. An example of the latter method was mentioned above (§7.2.1), and both will be encountered later (§9.1.2). They are really attempts to explain the nature of the inductive effect, and an important question which they raise is that of the relative importance of localisation and electrostatic phenomena in determining orientation and state of activation in electrophilic substitutions.

7.3 QUANTITATIVE CORRELATIONS OF SUBSTITUENT EFFECTS

The theories outlined above are concerned with the way in which substituents modify the reactivity of the aromatic nucleus. An alternative approach to the effects of substituents is provided by quantitative correlations of reactivity in which, at least to a certain extent, relative reactivities in different aromatic substitutions can be expressed by equations containing one term characteristic only of the substituent, and another term which depends only on the reaction.

Irrespective of the precision of these quantitative correlations, this approach is useful in emphasizing that relative rates depend on the nature of the reaction as well as of the aromatic compound.

The best-known equation of the type mentioned is, of course, Hammett's equation. It correlates, with considerable precision, rate and equilibrium constants for a large number of reactions occurring in the side chains of *m*- and *p*-substituted aromatic compounds, but fails badly for electrophilic substitution into the aromatic ring (except at *m*-positions) and for certain reactions in side chains in which there is considerable mesomeric interaction between the side chain and the ring during the course of reaction.[29] This failure arises because Hammett's original model reaction (the ionization of substituted benzoic acids) does not take account of the direct resonance interactions between a substituent and the site of reaction. This sort of interaction in the electrophilic substitutions of anisole is depicted in the following resonance structures, which show the transition state to be stabilized by direct resonance with the substituent:

There were two schools of thought concerning attempts to extend Hammett's treatment of substituent effects to electrophilic substitutions. It was felt by some that the effects of substituents in electrophilic aromatic substitutions were particularly susceptible to the specific demands of the reagent, and that the variability of the polarizibility effects, or direct resonance interactions, would render impossible any attempted correlation using a two-parameter equation.[29-30] This view was not universally accepted, for Pearson, Baxter and Martin[31] suggested that, by choosing a different model reaction, in which the direct resonance effects of substituents participated, an equation, formally similar to Hammett's equation, might be devised to correlate the rates of electrophilic aromatic and electrophilic side chain reactions. We shall now consider attempts which have been made to do this.

Aromatic reactivity: A. Theoretical background

7.3.1 Linear free energy relationships

Streitwieser[4a] pointed out that the correlation which exists between relative rates of reaction in deuterodeprotonation, nitration, and chlorination, and equilibrium constants for protonation in hydrofluoric acid amongst polynuclear hydrocarbons (cf. §6.2.3) constitutes a relationship of the Hammett type. The standard reaction is here the protonation equilibrium (for which ρ^* is unity by definition). For convenience he selected the 1-position of naphthalene, rather than a position in benzene as the reference position (for which σ^* is zero by definition), and by this means was able to evaluate ρ^*-values for the substitutions mentioned, and σ^*-values for positions in a number of hydrocarbons. The ρ^*-values (for protonation equilibria, 1; for deuterodeprotonation, 0·47; for nitration, 0·26; and for chlorination, 0·64) are taken to indicate how closely the transition states of these reactions resemble a σ-complex.

The more extensive problem of correlating substituent effects in electrophilic substitution by a two-parameter equation has been examined by Brown and his co-workers.[32] In order to define a new set of substituent constants, Brown chose as a model reaction the solvolysis of substituted dimethylphenylcarbinyl chlorides in 90% aq. acetone. In the case of p-substituted compounds, the transition state, represented by the following resonance structures, is stabilized by direct resonance interaction between the substituent and the site of reaction.

A plot against Hammett's σ-constants of the logarithms of the rate constants for the solvolysis of a series of m-substituted dimethylphenylcarbinyl chlorides, in which compounds direct resonance interaction with the substituent is not possible, yielded a reasonably straight line and gave a value for the reaction constant (ρ) of $-4\cdot54$. Using this value of the reaction constant, and with the data for the rates of solvolysis, a new set of substituent parameters (σ^+) was defined. The procedure described above for the definition of σ^+, was adopted for convenience so that values of σ^+ for *meta*-substituents were close to the corresponding values of σ defined by Hammett.

The suitability of the model reaction chosen by Brown has been criticised.[33] There are many side-chain reactions in which, during reaction, electron deficiencies arise at the site of reaction. The values of the substituent constants obtainable from these reactions would not agree with the values chosen for σ^+. At worst, if the solvolysis of sub-stituted benzyl chlorides in 50% aq. acetone had been chosen as the model reaction, σ^+_{p-Me} would have been -0.82 instead of the adopted value of -0.28. It is difficult to see how the choice of reaction was defended, save by pointing out that the variation in the values of the substituent constants, derivable from different reactions, were not systematically related to the values of the reaction constants (ρ);[32] such a relationship would have been expected if the importance of the stabi-lization of the transition-state by direct resonance increased with increasing values of the reaction constant.

The applicability of the two-parameter equation and the constants devised by Brown to electrophilic aromatic substitutions was tested by plotting values of the partial rate factors for a reaction against the appropriate substituent constants. It was maintained that such com-parisons yielded satisfactory linear correlations for the results of many electrophilic substitutions, the slopes of the correlations giving the values of the reaction constants.[32] If the existence of linear free energy relationships in electrophilic aromatic substitutions were not in dispute, the above procedure would suffice, and the precision of the correlation would measure the usefulness of the $\rho^+\sigma^+$ equation. However, a point at issue was whether the effect of a substituent could be represented by a constant, or whether its nature depended on the specific reaction. To investigate the effect of a particular substituent in different reactions, the values for the various reactions of the logarithms of the partial rate factors for the substituent were plotted against the ρ^+ values of the reactions. This procedure should show more readily whether the effect of a substituent depends on the reaction, in which case deviations from a linear relationship would occur. It was concluded that any variation in substituent effects was random, and not a function of electron demand by the electrophile.[32]

The use of Brown's equation ($\log_{10} k/k_0 = \rho^+\sigma^+$) with electrophilic substitutions in general has been fully discussed[32,34] and reference will be made later to its treatment of particular substituents in nitrations.

Aromatic reactivity: A. Theoretical background

With rates of nitration in sulphuric acid, Brown's σ^+-constants do not correlate well.[11]

To meet the point that the amount of resonance interaction in the transition state will be dependent upon the nature of the electrophile, Yukawa and Tsuno[33b] have put forward a modified equation with three parameters. The physical interpretation of such an equation is interesting, but it is not surprising that it correlates experimental data better than does the equation with two parameters.[34]

7.3.2 Reactivity and selectivity

Brown[32, 35] developed the 'selectivity relationship' before the introduction of σ^+ constants made possible correlations of aromatic reactivities following the Hammett model. The former, less direct approach to linear free-energy relationships was necessary because of lack of data at the time.

Brown noticed that the reactivities of toluene relative to benzene in aromatic substitutions were proportional to the ratios in which toluene underwent p- and m-substitutions. This point is illustrated in table 7.3.

TABLE 7.3   *Relative rates and isomer ratios for substitution* in benzene and toluene[32]*

Reaction	Relative rate k_T/k_B	Product distribution	
		% meta	% para
Bromination†	605	0·3	66·8
Chlorination‡	350	0·5	39·7
Benzoylation§	110	1·5	89·3
Nitration ‖	23	2·8	33·9
Mercuration¶	7·9	9·5	69·5
Isopropylation**	1·8	26·6	47·2

* 25 °C. † Br_2-AcOH-H_2O. ‡ Cl_2-AcOH. § Ph.COCl-$AlCl_3$. ‖ $AcONO_2$-Ac_2O.
¶ $Hg(OAc)_2$-$HClO_4$-AcOH. ** i-PrBr-$GaBr_3$-ArH.

The above proportionality can be expressed by the following equation, in which $f_{p-\mathrm{Me}}$ and $f_{m-\mathrm{Me}}$ are the partial rate factors for substitution at p- and m-positions respectively.*

$$\log_{10} f_{p-\mathrm{Me}} = b \log_{10}(f_{p-\mathrm{Me}}/f_{m-\mathrm{Me}}).$$

* de la Mare[30b] has pointed out the relationship between treatments of this kind and those having the general form of Ri and Eyring's electrostatic theory (§7.2.4).

By defining the selectivity factor (S_f) of a reaction in the following way:

$$S_f \equiv \log_{10}(f_{p-\mathrm{Me}}/f_{m-\mathrm{Me}}),$$

the previous equation can be written in the following form:

$$\log_{10} f_{p-\mathrm{Me}} = bS_f.$$

This equation is formally similar to the $\rho^+\sigma^+$ relationship,[36] S_f being related to ρ^+, and b being a parameter whose value depends on the values of σ^+ for *meta-* and *para*-methyl groups $[b = \sigma_p^+/(\sigma_p^+ - \sigma_m^+)]$.

The selectivity of an electrophile, measured by the extent to which it discriminated either between benzene and toluene, or between the *meta-* and *para*-positions in toluene, was considered to be related to its reactivity.[37] Thus, powerful electrophiles, of which the species operating in Friedel–Crafts' alkylation reactions were considered to be examples, would be less able to distinguish between compounds and positions than a weakly electrophilic reagent. The 'ultimate electrophilic species' would be entirely insensitive to the differences between compounds and positions, and would bring about reaction in the statistical ratio of the various sites for substitution available to it.[32] The idea has gained wide acceptance that the electrophiles operative in reactions which have low selectivity factors (S_f) or reaction constants (p^+), are intrinsically more reactive than the effective electrophiles in reactions which have higher values of these parameters. However, there are several aspects of this supposed relationship which merit discussion.

The selectivity relationship merely expresses the proportionality between intermolecular and intramolecular selectivities in electrophilic substitution, and it is not surprising that these quantities should be related. There are examples of related reactions in which connections between selectivity and reactivity have been demonstrated. For example, the ratio of the rates of reaction with the azide anion and water of the triphenylmethyl, diphenylmethyl and *tert*-butyl carbonium ions were $2 \cdot 8 \times 10^5$, $2 \cdot 4 \times 10^2$ and $3 \cdot 9$ respectively; the selectivities of the ions decrease as the reactivities increase.[38] The existence, under very restricted and closely related conditions, of a relationship between reactivity and selectivity in the reactions mentioned above, does not permit the assumption that a similar relationship holds over the wide range of different electrophilic aromatic substitutions. In these substitution reactions a difficulty arises in defining the concept of reactivity; it is not sufficient to assume that the reactivity of an electrophile is related to the selectivity that it exhibits in reaction, because it is the connection between these two quantities which is in question. The reactivity of a species could properly be expressed by the rate constant for its reaction with a standard compound under standard conditions. However, it is experimentally impossible to bring about a large number of electrophilic reactions under similar conditions, and furthermore there are few reactions for which the nature and concentration of the active electrophilic species are known. These considerations suggest that comparisons of the relative reactivities of electrophilic species are only possible under special circumstances, which are that the species under consideration should be closely related and that the reactions should proceed by similar mechanisms under similar conditions.

Nitration in sulphuric acid is a reaction for which the nature and concentrations of the electrophile, the nitronium ion, are well established. In these solutions compounds reacting one or two orders of magnitude faster than benzene do so at the rate of encounter of the aromatic molecules and the nitronium ion (§2.5). If there were a connection between selectivity and reactivity in electrophilic aromatic substitutions, then electrophiles such as those operating in mercuration and Friedel–Crafts' alkylation should be subject to control by encounter at a lower threshold of substrate reactivity than in nitration; this does not appear to occur.

The occurrence of a hydrogen isotope effect in an electrophilic substitution will certainly render nugatory any attempt to relate the reactivity of the electrophile with the effects of substituents. Such a situation occurs in mercuration in which the large isotope effect ($k_H/k_D = 6$) has been attributed to the weakness of the carbon–mercury bond relative to the carbon–hydrogen bond.[39] The following scheme has been formulated for the reaction, and the occurrence of the isotope effect indicates that the magnitudes of k_{-1} and k_2 are comparable:

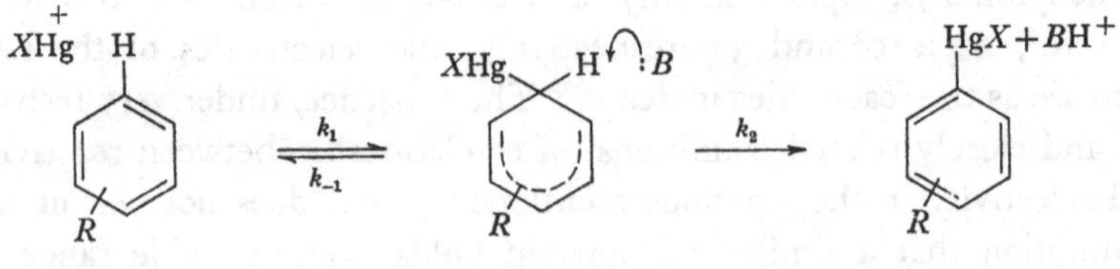

$$\overset{+}{X}Hg\ H \underset{k_{-1}}{\overset{k_1}{\rightleftharpoons}} XHg\ H\ :B \xrightarrow{k_2} HgX + BH^+$$

However, the existence of the Wheland intermediate is not demanded by the evidence, for if the attack of the electrophile and the loss of the proton were synchronous an isotope effect would also be expected. The donation of electrons from the substituent to the ring facilitates reaction with the incoming electrophile, but has an adverse effect on the ease of removal of the proton, which process is kinetically important in this reaction. Thus, because substituents have opposite effects on these two aspects of the reaction, mercuration is relatively insensitive to the effects of substituents ($\rho^+ = -4 \cdot 0$); it is not justifiable to conclude from the narrow spread of rates that the electrophile is very reactive.

In conclusion it might be said that quantitative linear correlations of aromatic reactivity using the $\sigma^+\rho^+$ relation can be drawn to a limited extent. Generally, the precisions of the correlations are not as good as those given by the Hammett equation for reactions in which direct resonance effects are not important. The assumption, implicit in the $\sigma^+\rho^+$ relation, that the direct resonance effects of substituents can be represented by a constant would receive its more severe test in the case of substituents, such as p-OMe and p-NMe$_2$, which by the mesomeric effect activate electrophilic substitution. However, because the data for these substituents are sparse and of poor quality, often involving lengthy extrapolations of results, this point is not proved.

The relationship of the selectivity of an electrophile to its reactivity is a separate issue, because the above quantitative correlations of reactivity can be used empirically, without accepting that they allow comment about the reactivity of electrophiles. There is no direct evidence for the view that differences in the selectivities of electrophiles are related in a simple way to their different reactivities. Indeed, it is difficult to grasp the meaning of comparisons attempted between electrophiles of very different structures, which bring about reaction under disparate conditions by different mechanisms.

REFERENCES

1. Ingold, C. K. (1953). *Structure and Mechanism in Organic Chemistry*. London: Bell.
2. Remick, A. E. (1949). *Electronic Interpretations of Organic Chemistry*, 2nd ed. New York: Wiley.
3. Murrell, J. N., Kettle, S. F. A. & Tedder, J. M. (1965). *Valence Theory*. New York: Wiley.
4. (a) Streitwieser, A. (1961). *Molecular Orbital Theory for Organic Chemists*. New York: Wiley.
   (b) Higasi, K., Baba, H. and Rembaum, A. (1965). *Quantum Organic Chemistry*. New York: Interscience.

References

5. Daudel, R. (1967). *Théorie quantique de la réactivité chimique*. Paris: Gauthier-Villars.
6. Wheland, G. W. (1955). *Resonance in Organic Chemistry*. New York: Wiley.
7. Dewar, M. J. S. (a) (1965). *Adv. Chem. Phys.* **8**, 65; (b) (1969). *The Molecular Orbital Theory of Organic Chemistry*. New York: McGraw-Hill.
8. Schofield, K. (1967). *Hetero-Aromatic Nitrogen Compounds: Pyrroles and Pyridines*. London: Butterworths.
9. Bell, R. P. (1959). *The Proton in Chemistry*. London: Methuen. Ritchie, C. D. and W. F. Sager. (1964). *Progress in Physical Organic Chemistry*, vol. 2, ed. S. G. Cohen, A. Streitwieser, &. R W. Taft. New York: Interscience.
10. (a) Coombes, R. G., Moodie, R. B. & Schofield, K. (1968). *J. chem. Soc.* B, p. 800.
    (b) Hoggett, J. G., Moodie, R. B. & Schofield, K. (1969). *J. chem. Soc.* p. 1.
11. Coombes, R. G., Crout, D. H. G., Hoggett, J. G., Moodie, R. B. & Schofield, K. (1970). *J. chem. Soc.* (B), p. 347.
12. Bell, R. P. & Ramsden, E. N. (1958). *J. chem. Soc.* p. 161.
    Bell, R. P. & De Maria, P. (1969). *J. chem. Soc.* B, p. 1057.
    Dubois, J.-E., Alcais, P. & Barbier, G. (1968). *Bull. Soc. chim. Fr.* pp. 605, 611.
    Dubois, J.-E., Uzan, R. & Alcais, P. (1968). *Bull. Soc. chim. Fr.* p. 617.
    Dubois, J.-E. & Uzan, R. (1968). *Bull. Soc. chim. Fr.* p. 3534.
13. de la Mare, P. B. D. & Ridd, J. H. (1959). *Aromatic Substitution; Nitration and Halogenation*. London: Butterworths.
14. Dewar, M. J. S. & Grisdale, P. J. (1962). *J. Am. chem. Soc.* **84**, (a) p. 3539, (b) p. 3548.
15. Branch, G. E. K. & Calvin, M. (1945). *The Theory of Organic Chemistry*. New York: Prentice-Hall.
16. Brown, D. A. & Dewar, M. J. S. (1953). *J. chem. Soc.* p. 2406.
17. Roberts, J. D. with (a) Moreland, W. T. (1953). *J. Am. chem. Soc.* **75**, 2167; (b) with Clement, R. A. & Drysdale, J. J. (1951). **73**, 2181.
18. Robinson, R. (1932). *Two Lectures on an Outline of an Electrochemical (Electronic) Theory of the Course of Organic Reactions*. London: Institute of Chemistry.
19. Waters, W. A. (1948). *J. chem. Soc.* p. 727.
20. Wheland, G. W. & Pauling, L. (1935). *J. Am. chem. Soc.* **57**, 2086.
21. Greenwood, H. H. & McWeeny, R. (1966). *Advances in Physical Organic Chemistry*, vol. 4, ed. V. Gold. London: Academic Press.
22. Wheland, G. W. (1942). *J. Am. chem. Soc.* **64**, 900.
23. Dewar, M. J. S. (a) (1952). *J. Am. chem. Soc.* **74**, 3341; (b) with Mole, T. and Warford, E. W. T. (1956). *J. chem. Soc.* p. 3581; (c) with Thompson, C. C. (1965). *J. Am. chem. Soc.* **87**, 4414.
24. Mason, S. F. (1958). *J. chem. Soc.* p. 4329; (1959). p. 1233.
25. Ridd, J. H. (1963). *Physical Methods in Heterocyclic Chemistry*, vol. 1, ed. A. R. Katritzky. London: Academic Press.
26. Hudson, R. F. & Klopman, G. (1967). *Tetrahedron Lett.* p. 1103.
    Klopman, G. (1968). *J. Am. chem. Soc.* **90**, 223.
27. Ri, T. & Eyring, H. (1940). *J. chem. Phys.* **8**, 433.

28. Kenner, G. W. (1946). *Proc. R. Soc.* A **185**, 119.
29. Roberts, J. D., Sanford, J. K., Sixma, F. L. J., Cerfontain, H. & Zagt, R. (1954). *J. Am. chem. Soc.* **76**, 4525.
30. (a) Swain, C. G. & Langsdorf, W. P. (1951). *J. Am. chem. Soc.* **73**, 2813.
    (b) de la Mare, P. B. D. (1954). *J. chem. Soc.* p. 4450.
31. Pearson, D. E., Baxter, J. F. & Martin, J. G. (1952). *J. org. Chem.* **17**, 1511.
32. Brown, H. C. & Stock, L. M. (1962). *J. Am. chem. Soc.* **84**, 3298.
    Stock, L. M. & Brown, H. C. (1963). In *Advances in Physical Organic Chemistry*, vol. 1, ed. V. Gold. London: Academic Press.
33. (a) van Bekkum, H., Verkade, P. E. & Wepster, B. M. (1959). *Recl Trav. chim. Pays-Bas Belg.* **78**, 815.
    (b) Yukawa, Y. & Tsuno, Y. (1959). *Bull. chem. Soc. Japan* **32**, 971.
34. Norman, R. O. C. & Taylor, R. (1965). *Electrophilic Substitution in Benzenoid Compounds.* London: Elsevier.
35. Brown, H. C. & McGary, C. W. (1955). *J. Am. chem. Soc.* **77**, 2300.
36. McGary, C. W., Okamoto, Y. & Brown, H. C. (1955). *J. Am. chem. Soc.* **77**, 3037.
37. Brown, H. C. & Nelson, K. L. (1953). *J. Am. chem. Soc.* **75**, 6292.
38. Swain, C. G., Scott, C. B. & Lohmann, K. H. (1953). *J. Am. chem. Soc.* **75**, 136.
39. Kresge, A. J. & Brennan, J. F. (1963). *Proc. chem. Soc.* 215.
    Perrin, C. & Westheimer, F. H. (1963). *J. Am. chem. Soc.* **85**, 2773.

8 Nitration and aromatic reactivity: B. The nitration of bases

8.1 INTRODUCTION

Nitration is almost always carried out under acidic conditions. If the compound being nitrated is basic, the problem arises of deciding whether the free base or its conjugate acid is being nitrated, or if both of these species are reacting.

Nitration may or may not involve the predominant form of the substrate. In the latter case, if the predominant form is the conjugate acid, the observed second-order rate constant can be corrected to give one ($k_2$fb.) appropriate to the reacting free base. With a reaction of the form

$$SH^+ \rightleftharpoons S + H^+$$

$$S + NO_2^+ \longrightarrow \text{product},$$

we have the following relationships:

$$\text{rate} = k_2\text{obs.}[S]_{\text{stoich}}[HNO_3]$$

$$= k_2\text{fb.}[S][HNO_3],$$

$$k_2\text{fb.} = k_2\text{obs.}[S]_{\text{stoich}}/[S]$$

$$= k_2\text{obs.}(1+I).$$

The ionisation ratio ($I = [SH^+]/[S]$) can be calculated from a knowledge of the acidity function (h_x) followed by the substrate, and the acidity constant of the conjugate acid. Thus, when $I \gg 1$:

$$\log_{10}k_2\text{fb.} = \log_{10}k_2\text{obs.} + pK_a - H_x.$$

Several criteria have been used in identifying the reacting species in these reactions. These are:

(1) The variation of the observed second-order rate constant with acidity.

(2) Comparison of the rate of nitration of the base with that of a necessarily cationic derivative.

(3) Comparison of the observed rate of nitration with the calculated encounter rate.

146

(4) Determination of the Arrhenius parameters.

(5) Consideration of the orientation of substitution.

These criteria, which have been applied almost solely to nitrations in sulphuric acid, will now be discussed.

8.2 RECOGNITION OF THE REACTING SPECIES

8.2.1 *The variation of the observed second-order rate constant with acidity*

In moderately concentrated (60–85 %) *aqueous sulphuric acid.* For neutral compounds such as those listed in table 2.5, plots of the logarithm of the observed second-order rate constant for nitration at 25 °C against the percentage of sulphuric acid have similar slopes, $d(\log_{10} k_2 \text{obs.})/d(\% \text{ H}_2\text{SO}_4) \sim 0\cdot3$–$0\cdot4$, the higher values being obtained when studies were made in the upper range of acidity in question. The same is true for quaternary cations (table 8.1, section A). The observation has also been made that plots of $\log_{10} k_2$obs. *v.* $-(H_R + \log_{10} a_{\text{H}_2\text{O}})$ for both neutral molecules and quaternary cations are usually linear, with slopes close to unity. The explanation preferred[1] was that nitric acid in aqueous sulphuric acid is present predominantly as the monohydrate, so that the situation under consideration can be written as follows:

$$\text{HNO}_3 . \text{H}_2\text{O} + \text{H}^+ \overset{K}{\rightleftharpoons} \text{NO}_2^+ + 2\text{H}_2\text{O},$$

$$\text{ArH} + \text{NO}_2^+ \overset{k}{\longrightarrow} \text{products.}$$

Use of the Brönsted rate equation gives the following expression:

$$\text{rate} = k a_{\text{NO}_2^+} a_{\text{ArH}} / f_{\ddagger}.$$

Remembering that the observed second-order rate constant is merely the rate divided by the product of the stoichiometric concentrations of aromatic compound and nitric acid, the following relationship can be derived

$$k_2 \text{obs.} = kK f_{\text{ArH}} f_{\text{HNO}_3} a_{\text{H}+} / a_{\text{H}_2\text{O}}^2 f_{\ddagger}$$

and since by definition

$$h_R = a_{\text{H}+} f_{\text{ROH}} / f_{R+} a_{\text{H}_2\text{O}},$$

it follows that

$$\log_{10} k_2 \text{obs.} = -(H_R + \log_{10} a_{\text{H}_2\text{O}}) + \log_{10}(kK f_{R+} f_{\text{HNO}_3} f_{\text{ArH}} / f_{\text{ROH}} f_{\ddagger}).$$

TABLE 8.1 *The acidity dependence and Arrhenius parameters for the nitration of some cations in aqueous sulphuric acid*

Compound	Temp. /°C	Slope			Arrhenius parameters			Ref.‡
		H_2SO_4/%	*	†	H_2SO_4/%	E/kJ mol⁻¹	$\log_{10}$ (A/l mol⁻¹ s⁻¹)	
Section A								
Benzyltrimethylammonium	25·1	74·5–81·5	0·38	1·02	70·0	79	9·1	·
2-Methoxyisoquinolinium	25·0	76·4–83·1	0·38	1·01	83·1	65	8·4	·
2-Methylisoquinolinium	25·0	76·4–83·7	0·42	1·12	81·3	62	8·4	·
1-Methylquinolinium	25·0	79·6–83·7	0·38	1·00	81·3	75	8·8	·
2-Phenethyltrimethylammonium	25·1	63·4–76·5	0·35	1·09	70·0	70	10·9	·
2-Phenethyltrimethylammonium					68·3	70	10·1	·
3-Phenpropyltrimethylammonium	25·1	61·0–68·3	0·32	1·02	68·3	69	11·0	·
Phenyltrimethylammonium	25·0	82·0–87·5	0·38	0·90	87·5	56	8·0	·
Phenyltrimethylstibonium§	25·0	75·9–82·1	0·40	1·08	·	·	·	·
p-Tolyltrimethylammonium	25·0	75·8–81·7	0·40	1·08	80·0	57	10·1	·
Section B								
2-Methoxycinnolinium	80	75·2–81·5	0·28				·	2
2-Methylcinnolinium	80	77·0–81·2	0·26		81·2	80	8·7	6
2-Methylisoquinolinium	80	64·4–73·4	0·26		·	·	·	

Section C

Compound	Temp.	% range						
Anilinium‖	25·0	82·0–86·5	0·39	0·97	·	·	·	·
Benzamidium	25·0	81·2–84·9	0·38	0·98	81·2	62	9·6	·
Benzylammonium‖	25·0	78·7–80·0	0·28	0·75	·	·	·	·
1-Hydroxy-2,6-dimethoxypyridinium	25·0	80·0–84·8	0·39	1·03	84·8	61	6·9	·
2-Hydroxyisoquinolinium	25·0	76·4–83·1	0·37	0·98	83·1	62	8·3	·
1-Hydroxy-2-phenylpyridinium	25·0	74·7–78·6	0·41	1·10	·	·	·	·
Imidazolium§‖	25·0	83·8–86·8	0·27	0·88	·	·	·	7
Isoquinolinium	25·0	78·1–83·7	0·40	1·05	81·3	59	7·7	·
Phenyldimethylhydroxylammonium	25·0	84·4–86·4	0·37	0·92	·	·	·	·
2-Phenylpyridinium	25·0	77·1–80·9	0·39	1·02	·	·	·	·
Quinolinium§	25·0	79·6–83·7	0·36	0·95	81·3	69	8·3	·

Section D

Compound	Temp.	% range						
2-Cinnolinium	80	77·0–83·0	0·26	·	81·1	86	9·2	6
1-Hydroxyquinolinium¶	0	82·0–85·0	0·40	·	82·0	68	7·8	·
Isoquinolinium	80	67·7–73·0	0·26	·	·	·	·	6
Pyrazolium§‖	80·8	83·7–87·5	0·22	·	·	·	·	7

* $d(\log_{10} k_2)/d(\% \ H_2SO_4)$. † $d(\log_{10} k_2)/d[-(H_R + \log_{10} a_{H_2O})]$. ‡ Where no reference is given see table 2.5. § Cf. table 8.3. ‖ Rates were measured at only two acidities. ¶ For nitration at the 5- and 8-positions only.

Aromatic reactivity: B. Bases

Provided that the ratio of activity coefficients is invariant over the range of acidity concerned, a linear relationship with unit slope between $\log_{10} k_2$ obs. and $-(H_R + \log_{10} a_{H_2O})$ is to be expected. However, there is no way in which the assumption concerning the activity coefficients can be tested. Another serious objection to this explanation has been revealed by later work. In the few cases where the necessary measurements have been made it has been observed that plots of $\log_{10} k_2$ obs. v. % H_2SO_4 become less steep with increasing temperature[2] (table 8.1, section B). Studies of the temperature dependence of $(-H_R)$ show[3] that this function exhibits the opposite trend, so that the linear correlation with unit slope between $\log_{10} k_2$ obs. and $(-H_R + \log_{10} a_{H_2O})$ is almost certainly restricted to rate constants determined at or near 25 °C. Some uncertainty exists about the validity of the H_R scale which has been determined for temperatures other than 25 °C; Shapiro[4] has criticised the work on the grounds that the overlap between adjacent indicators used in the measurements was inadequate, and that agreement was not observed between values of H_R at 25 °C obtained from Arnett's data[3] and those of Deno and Stein.[5] In view of all these considerations it seems best to regard the correlation in question as an empirical one, and to use it only for rate constants determined at or near 25 °C.

For basic compounds which are predominantly protonated in the media in which nitrations are conducted, similar slopes,

$$d(\log_{10} k_2 \text{obs.})/d(\% \ H_2SO_4) \sim 0.3 - 0.4$$

and $d(\log k_2 \text{obs.})/d[-(H_R + \log a_{H_2O})] \sim 1$, are to be expected if the conjugate acid is the species undergoing nitration. This is the case with the compounds listed in table 8.1, section C. At temperatures far removed from 25 °C, different slopes may be obtained (table 8.1, section D) which may still be shown to indicate that nitration involves the conjugate acid by application of the second criterion (§8.2.2).

The unprotonated form of the base will generally be the more reactive toward nitration, and it is possible that nitration will occur through the free base even when its concentration is very small compared with that of the conjugate acid. In such cases substantially lower slopes are to be expected, because the proportion of free base falls with increasing acidity. Examples of compounds exhibiting low slopes which are attributable to nitration through the minority free base are given in table 8.2. The observed second-order rate constants can be corrected to give values relevant to the free base using the equation given above

(§8.1), and ideally the corrected rate profile should have a slope similar to those given by substrates which are nitrated *via* the dominant species. In practice this is rarely achieved. One difficulty in applying the equation arises from the fact that the acidity function followed by the compound under study is known only in a few cases. Obviously, different values for the slope will be obtained if different acidity functions are used in the calculation. For example, in the case of 2,6-lutidine 1-oxide the values of

TABLE 8.2   *The acidity dependence of rates of nitration of some free bases in sulphuric acid*

Compound	Temp.°C	$H_2SO_4/\%$	Slope *	Slope †	Reference
		(< 90%)			
Acetophenone	25	75·5–85·1	0·27	0·76	8
	25	75·5–85·1	0·36‡	0·95‡	8
Benzoic acid	25	78·0–81·4	0·34	0·90	8
	25	78·0–81·4	0·38‡	1·0‡	8
1,5-Dimethyl-2-pyridone	39·5	71·4–78·4	0·19	0·56	9
	39·5	71·4–78·4	.	0·86§	9
2,6-Lutidine 1-oxide	25 ‖	78·2–84·3	0·23	0·60	10
	25 ‖	78·2–84·3	0·31§	0·80§	10
	25 ‖	78·2–84·3	0·40¶	1·04¶	10
3-Methyl-2-pyridone	34·5	75·0–80·9	0·14	0·39	9
	34·5	75·0–80·9	.	0·65§	9
5-Methyl-2-pyridone	35	74·1–80·1	0·17	0·47	9
	35	74·1–80·1	.	0·65§	9
4-Pyridone	157·5	72·8–80·1	0·16	0·41	9
	157·5	72·8–80·1	.	0·66§	9
	157·5	72·8–80·1	.	0·86¶	9
Quinoline 1-oxide**	25 ‖	55–82	0·16	0·46	2
	25 ‖	55–82	0·23§	0·67	2
	25 ‖	55–82	0·30¶	0·88	2
		(> 90%)			
Acetophenone	25	91·4–98·1	−0·23	.	8
2-Chloro-4-nitroaniline	25	93·8–98·0	−0·23	.	11
1,5-Dimethyl-2-pyridone	29	92·5–97·9	−0·23	.	9
3-Methyl-2-pyridone	31	91·8–97·8	−0·24	.	9
5-Methyl-2-pyridone	31·5	90·8–96·8	−0·21	.	9
p-Nitroaniline	24·9	93·8–98·0	−0·17	.	11

* $d(\log_{10}k)/(d\%\ H_2SO_4)$.   † $d(\log_{10}k)/d[-(H_R+\log_{10}a_{H_2O})]$.
‡ Corrected from the ionisation ratio.   § Corrected by h_A/K_A.
‖ Data at 25 °C calculated from the Arrhenius parameters.
¶ Corrected by h_0/K_A.
** For nitration at the 4-position only.

the quantity $d(\log_{10} k_2 \text{fb.})/d(\% \text{ H}_2\text{SO}_4)$ at 25 °C obtained using H_A and H_0 are 0·31 and 0·40 respectively (table 8.2.)

Further problems arise if measurements of the rate of nitration have been made at temperatures other than 25 °C; under these circumstances two procedures are feasible. The first is discussed in §8.2.2 below. In the second the rate profile for the compound under investigation is corrected to 25 °C by use of the Arrhenius parameters, and then further corrected for protonation to give the calculated value of $\log_{10} k_2 \text{fb.}$ at 25 °C, and thus the calculated rate profile for the free base at 25 °C. The obvious disadvantage is the inaccuracy which arises from the Arrhenius extrapolation, and the fact that, as mentioned above, it is not always known which acidity functions are appropriate.

The most accurate method of deriving $k_2 \text{fb.}$ from $k_2 \text{obs.}$ is to use the equation $k_2 \text{fb.} = k_2 \text{obs.}(1 + I)$; the ionisation ratio of the compound under study being determined directly at the required acidity and temperature. In the cases where the temperature at which rates are measured is not 25 °C the way in which $k_2 \text{fb.}$ depends upon acidity will be given correctly, but again there will remain the difficulty that the slope to be expected at this temperature other than 25 °C is not known.

In concentrated sulphuric acid. The way in which the rate of nitration of some non-basic compounds depends upon acidity in the region above that of maximum rate ($\sim$ 90% sulphuric acid) has been discussed (§2.3.2). Cationic species behave similarly (table 2.4, fig. 2.1).

Values of $d(\log_{10} k_2 \text{obs.})/d(\% \text{H}_2\text{SO}_4)$ at acidities greater than 90% sulphuric acid for some free bases and cations are given in table 8.2, and in table 8.3. Curvature of the rate profile often makes accurate estimation of the slope difficult. However, in general the values of this quantity for compounds reacting *via* the free bases are larger than those given by cations.

Acidity function plots have been used in another way.[9] In 65–90% sulphuric acid the concentration of the nitronium ion is not equal to the concentration of nitric acid; the quantity

$$k_2^* = k_2 \text{obs.}[\text{HNO}_3]_{\text{stoich}}/[\text{NO}_2^+]$$

has been used to allow for incomplete ionisation of nitric acid in this region. Values of k_2^* for a number of compounds have been calculated up to 95% sulphuric acid, assuming that nitric acid is half-ionised in

88 % sulphuric acid and that the ionisation follows $-(H_R + \log_{10} a_{H_2O})$. From the following schemes

$$[S]_{\text{stoich}} + [HNO_3]_{\text{stoich}} \xrightarrow{k_2 \text{ obs}} \text{product},$$

$$[S]_{\text{stoich}} + [NO_2^+] \xrightarrow{k_2^*} \text{product},$$

$$[S] + [NO_2^+] \xrightarrow{k_2} \text{product},$$

we have $k_2^* = k_2 K_a/(K_a + h_x)$, where K_a is the ionisation constant of the conjugate acid SH^+, and h_x the acidity function appropriate to the ionisation of this acid. When $h_x \gg K_a$, then

$$\log_{10} k_2^* = H_x + \text{constant}.$$

Similarly, if nitration involves the conjugate acid, $k_2^* = k_2 h_x/(K_a + h_x)$, and when $h_x \gg K_a$
$$\log_{10} k_2^* = \text{constant}.$$

Thus, when nitration involves a free base, a plot of $\log_{10} k_2^*$ against $-H_0$ should have a slope of a (where $a = H_x/H_0$), but if it involves the conjugate acid the slope should be zero.

It is found in practice that for a number of compounds reacting *via* the predominant species an almost horizontal plot is obtained. For compounds presumed to be nitrated *via* the free bases, such as 2,6-lutidine 1-oxide and 3- and 5-methyl-2-pyridone, slopes of approximately unity are obtained.[9] Since this type of plot allows for the incomplete ionisation of nitric acid, it can be used at higher acidities than plots using $-(H_R + \log_{10} a_{H_2O})$ which break down when the condition $h_R/a_{H_2O} \gg K$ is no longer true.

8.2.2 *Comparison of the rate of nitration of the base with that of a necessarily cationic derivative*

This method is exemplified by its application to quinoline,[1] isoquinoline,[1] cinnoline,[6] and isoquinoline 2-oxide,[10] which are nitrated as their conjugate acids. The rate profiles for these compounds and their N- or O-methyl perchlorates show closely parallel dependences upon acidity (fig. 2.4). Quaternisation had in each case only a small effect upon the rate, making the criterion a very reliable one. It has the additional advantage of being applicable at any temperature for which kinetic measurements can be made (table 8.1, sections B and D).

The criterion has also been applied qualitatively. Thus, 2,4,6-collidine and 1,2,4,6-tetramethylpyridinium cation can be nitrated under identical conditions, suggesting that the protonated collidine is probably the form being nitrated in the first case.[12] Rate profiles for the nitration of these compounds in oleum support this conclusion.[13] In the case of pyridine 1-oxide, conditions which effected nitration with a half-life of 20 min gave none of the nitro-compound from 1-methoxypyridinium in 144 h; the free base is presumably being nitrated in the first case.[10] A similar observation was made with 2,6-lutidine 1-oxide and its quaternary derivative.[10] In the case of quinoline 1-oxide the behaviour of 1-methoxyquinolinium is consistent with the view that 5- and 8-nitration of the former involves the conjugate acid.[2]

8.2.3 *Comparison of the observed rate of nitration with the calculated encounter rate*

The rate constant for the collision of two species is given by the following approximate formula, and corresponds to a maximum value of the rate constant of reaction (§§2.5, 3.3):

$$k_2 \text{ enc.} = 8RT/3\eta.$$

For a base the stoichiometric second-order rate constant which should be observed, under conditions where ionisation to the nitronium ion is virtually complete, namely $> 90\% \ H_2SO_4$, if nitration were limited to the free base and occurred at every encounter with a nitronium ion, would be:

$$k_2 \text{ calc.} = k_2 \text{ enc. } [S]/[S]_{\text{stoich}}.$$

If the observed second-order rate constant is greater than k_2 calc., reaction *via* the free base is precluded. If $k_2$obs. is less than $k_2$calc., reaction *via* the conjugate acid or the free base is possible.[7] The first compound reported to be nitrated *via* its conjugate acid, and yet to have k_2 calc. $> k_2$ obs. at the acidities concerned, was pyrazole;[7] other examples are mentioned later (§§9.3; 10.4.2).

In applying this criterion, k_2 obs. must be compared with k_2 calc. for the same temperature. In general this entails knowledge of the temperature dependence of the relevant acidity function and of the ionisation constant. The latter factor has sometimes been allowed for (as in the calculation[13] of k_2 calc. for the nitration of 2,4,6-trimethylpyridine in 98% sulphuric acid at 80 °C) by using the approximate relationship,[14] $-d(pK_a)/dT = (pK_a - 0.9)/T$.

In the absence of accurate knowledge of $d(pK_a)/dT$ and dH_x/dT, the experimental rate data must be extrapolated to 25 °C.

8.2.4   *Determination of Arrhenius parameters*

Data are given in tables 8.1, 8.3 and 8.4.

In principle the use of the entropy of activation as a criterion is straightforward. The electrostatic contribution to this quantity, $\Delta S_{el}^{\ddagger}$, for a reaction between two cations is predicted from simple electrostatic theory to be less than that for a reaction between an ion and a neutral molecule. If the reactions are otherwise similar, the *overall* entropies of activation can be expected to differ in the same way:

$$\Delta S_{el}^{\ddagger} = -\frac{LZ_AZ_Be^2}{4\pi\epsilon r}\left(\frac{\partial \ln \epsilon}{\partial T}\right)_p.$$

In this formula,[15] which is presented here in the rationalised s.i. form, ϵ is the permittivity of the medium (the product of the dielectric constant and the permittivity of a vacuum) and r is the distance of separation of the charges in the transition state. For two monocations with $r =$ 0·2 nm in water at 25 °C (for which medium $\epsilon = 7·0 \times 10^{-10}$ kg^{-1} m^{-3} s^4 A^2 and $(\partial \ln\epsilon/\partial T)_p = -0·0046$ K^{-1}), $\Delta S_{el}^{\ddagger} \simeq -40$ JK^{-1} mol^{-1}. This then is the magnitude of the difference in entropies of activation to be expected for the elementary reaction in water at 25 °C between the nitronium ion and a neutral molecule on the one hand, and the nitronium ion and a cation on the other. This is equivalent to the expectation that the term $\log_{10}(A/\text{l mol}^{-1}\text{ s}^{-1})$, where A is the Arrhenius pre-exponential factor, would be greater by about 2 for reactions of neutral molecules than for reactions of cationic substrates with the nitronium ion. However, there are (at least) five serious drawbacks to the use of this expectation as a criterion to decide whether a positively charged conjugate acid or a neutral free base is the species undergoing nitration. These are:

(1) The error in the determination of $\Delta S^{\ddagger}$ is usually about ± 6 JK^{-1} mol^{-1} (corresponding to an error in $\log_{10}(A/\text{l mol}^{-1}\text{ s}^{-1})$ of $\pm 0·3$).

(2) The geometry of the transition state is always unknown, so that r can only be guessed. This is probably the least serious of the problems.

(3) It is necessary to use the bulk permittivity of the solvent in the equation, instead of the unknown but more correct 'effective' permittivity of the medium between the charges in the transition state.

(4) Even the bulk permittivities of aqueous sulphuric acid solutions are unknown.

(5) The observed entropy of activation for nitration through a

TABLE 8.3 *The acidity dependence of rate, and the Arrhenius parameters for the nitration of some cations in concentrated sulphuric acid*

Compound	Temp/°C	H_2SO_4/%	Slope†	H_2SO_4/%	$E/\text{kJ mol}^{-1}$	$\log_{10}(A/\text{l mol}^{-1}\,\text{s}^{-1})$	Ref.
Anilinium‡	25	92·4–100	−0·07				18
p-Chlorophenyltrimethyl-ammonium	25	92·7–98·4	−0·07	90·3	68	9·6	19
Imidazolium‡	25	93·8–99·3	−0·11	94·7	71	10·0	19
2-Methoxy-3-methyl-pyridinium	25·9	90·6–95·8	−0·09	98·7	73	10·0	19
Phenyltrimethylammonium‡	25	91·1–98·9	−0·06	95·3	65		7
Pyrazolium‡	80·8	92–99	−0·11	99·6	68	9·8	9
Quinolinium‡	25·3	91·9–98·9	−0·11	98·6	73	7·8	19, 20
				98·0	63	9·0	7
							21

* See also tables 2.3 and 2.4. † $d(\log_{10} k_2)/d(\%\,H_2SO_4)$. ‡ Cf. table 8.1.

minority free base includes a contribution from the entropy of acidic dissociation of the conjugate acid in the medium in question.

One approach to problems (3) and (4) is to use the experimental difference in entropies of activation of neutral molecules and cationic substrates, considering initially only cases where the majority species is known to be the one undergoing nitration. Such comparisons must be made for closely similar media, because entropies of activation are known to be very sensitive to the concentration of the acid.[16] The largest amount of information relates to 80–83 % H_2SO_4. In such media it is found[2, 17] that the value of $\log_{10}(A/\mathrm{l\,mol^{-1}s^{-1}})$ falls between 7·7 and 9·6 whether the species undergoing nitration is neutral or cationic. This suggests that the quantity $\epsilon^{-1}(\partial \ln \epsilon/\partial T)_p$ is small for these media, and finally demolishes the idea of using the entropy of activation as a criterion in the way described above.

An alternative approach[2] is to assume, in the light of the experimental evidence just mentioned, that the reactions of cations and neutral molecules have similar values of $\Delta S^{\pm}$ (or, equivalently, of $\log_{10} (A/\mathrm{l\ mol^{-1}\ s^{-1}})$), and to try to calculate the difference which would arise from the fact that the observed entropy of activation for a minority free base includes a contribution from the acidic dissociation of the conjugate acid in the medium in question (see (5) above). Consider the two following reaction schemes: one (primed symbols) represents nitration *via* the free base, the other the normal nitration of a non-basic majority species (unprimed symbols):

$$HNO_3 + H^+ \rightleftharpoons NO_2^+ + H_2O \qquad\qquad HNO_3 + H^+ \rightleftharpoons NO_2^+ + H_2O$$

$$S'H^+ \xrightleftharpoons{K_a} S' + H^+ \qquad\qquad\qquad S + NO_2^+ \xrightarrow{k_2} products$$

$$S' + NO_2^+ \xrightarrow{k_2'} products$$

For the first scheme the Brönsted treatment gives:

$$\text{rate} = k_2'\, a_S'\, a_{NO_2^+}/f_{\ddagger}'$$

$$= [S']_{\text{stoich}}\, [NO_2^+]\, k_2'\, K_a f_{S'H^+}\, f_{NO_2^+}/a_{H^+}\, f_{\ddagger}'.$$

Thus,

$$k_{\text{obs}}' = \text{rate}/[S']_{\text{stoich}}\, [HNO_3]_{\text{stoich}}$$

$$= k_2'\, K_a f_{S'H^+}\, f_{NO_2^+}\, [NO_2^+]/a_{H^+}\, f_{\ddagger}'\, [HNO_3]_{\text{stoich}}.$$

From a similar treatment of the second scheme we then have:

$$k_{\text{obs}}'/k_{\text{obs}} = k_2'\, K_a f_{\ddagger}\, f_{S'H^+}/k_2\, f_{\ddagger}'\, f_S\, a_{H^+}.$$

If the protonation of S′ follows an acidity function h_x, and if we assume that under the experimental conditions $f_S/f_{\ddagger} = f_{S'}/f'_{\ddagger}$, then

$$k'_{\mathrm{obs}}/k_{\mathrm{obs}} = k'_2\,K_a/k_2 h_x.$$

By definition $\log_{10}A = \log_{10}k - T^{-1}[d(\log_{10}k)/d(T^{-1})]$, and if we assume that $\log_{10}A'_2 = \log_{10}A_2$, we can derive the following equation:

$$\log_{10}A'_{\mathrm{obs}} = \log_{10}A_{\mathrm{obs}} - \mathrm{p}K_a + H_x - T^{-1}[d(\log_{10}K_a)/d(T^{-1})]$$
$$- T^{-1}[dH_x/d(T^{-1})].$$

From van't Hoff's equation we then have

$$\log_{10}A'_{\mathrm{obs}} = \log_{10}A_{\mathrm{obs}} + \Delta S^{\circ}/2{\cdot}303R + H_x - T^{-1}[dH_x/d(T^{-1})].$$

From the equation $k'_{\mathrm{obs}}/k_{\mathrm{obs}} = k'_2\,K_a/k_2 h_x$ (see above) we have

$$d[\ln(k'_{\mathrm{obs}}/k_{\mathrm{obs}})]/d(T^{-1}) = d[\ln(k'_2/k_2)]/d(T^{-1}) + d[\ln(K_A/h_x)]/d(T^{-1}).$$

From the Arrhenius equation and the assumption that

$$\log_{10}A'_2 = \log_{10}A_2$$

it follows that $\quad \ln(k'_2/k_2) = -(E'_2 - E_2)/RT,$

so that $\quad d[\ln(k'_2/k_2)]/d(1/T) = -(E'_2 - E_2)/R$

$$= T\ln(k'_2/k_2)$$

$$= T\ln(k'_{\mathrm{obs}}h_x/k_{\mathrm{obs}}K_a).$$

Now $d(\ln k)/d(T^{-1}) = -E/R$, so that

$$-E'_{\mathrm{obs}}/R + E_{\mathrm{obs}}/R = T\ln(k'_{\mathrm{obs}}h_x/k_{\mathrm{obs}}K_a) + d[\ln(K_a/h_x)]/d(T^{-1}),$$

and finally:

$$E'_{\mathrm{obs}} = E_{\mathrm{obs}} - 2{\cdot}303RT\,(\log_{10}k'_{\mathrm{obs}} - \log_{10}k_{\mathrm{obs}} - H_x + \mathrm{p}K_a) + \Delta H^{\circ}$$
$$- 2{\cdot}303R\,[dH_x/d(T^{-1})].$$

These equations, relating A'_{obs} to A_{obs}, and E'_{obs} to E_{obs}, show that A'_{obs} and E'_{obs} can be calculated for a reaction proceeding through the equilibrium concentration of a free base if the thermodynamic quantities relating to the ionisation of the base, and the appropriate acidity function and its temperature coefficient are known (or alternatively, if the ionisation ratio and its temperature coefficient are known under the appropriate conditions for the base.[8])

These arguments were originally applied to the 4-nitration of 2,6-lutidine 1-oxide and quinoline 1-oxide, and use of the data available for the temperature coefficient of H_0 led to the conclusion that the simple mechanism, which supposed nitration to proceed *via* the small equilibrium concentration of these bases, was not tenable.[2]

Similar difficulties arise in the nitrations of 2-chloro-4-nitroaniline and *p*-nitroaniline.[11] Consideration of the rate profiles and orientation of nitration (§8.2.5) in these compounds suggests that nitration involves the free bases. However, the concentrations of the latter are so small as to imply that if they are involved reaction between the amines and the nitronium ion must occur upon encounter; that being so, the observed activation energies appear to be too high. The activation energy for the simple nitration of the free base in the case of *p*-nitroaniline was calculated from the following equation:

$$E = -2 \cdot 303 R[d(\log_{10} k_2 \mathrm{fb.})/d(T^{-1})] + \Delta H° - 2 \cdot 303 R[dH_0/d(T^{-1})].$$

The first term, the apparent activation energy of the encounter reaction, was evaluated from the temperature coefficient of the viscosity of sulphuric acid.

Because of these difficulties, special mechanisms were proposed for the 4-nitrations of 2,6-lutidine 1-oxide and quinoline 1-oxide,[2] and for the nitration of the weakly basic anilines.[11] However, recent re-measurements of the temperature coefficient of H_0, and use of the new values in the above calculations reconciles experimental and calculated activation parameters and so removes difficulties in the way of accepting the mechanisms of nitration as involving the very small equilibrium concentrations of the free bases.[4] Despite this resolution of the difficulty some problems about these reactions do remain, especially when the very short life times of the molecules of unprotonated amines in nitration solutions are considered.[11]

For the nitration of the very weak base, acetophenone, there is reasonable agreement between observed and calculated activation parameters, and there is no doubt that nitration of the free base occurs at acidities below that of maximum rate. In this case the equilibrium concentration of free base is much greater than in the examples just discussed and there is no question of reaction upon encounter.[8]

8.2.5 *Consideration of the orientation of substitution*

Orientation is an important factor to be considered in recognising both changes in the effective electrophile and in the nature of the aromatic substrate. Cases of the former type, which will be met at several places

TABLE 8.4   *The Arrhenius parameters for the nitration of some free bases in sulphuric acid*

Compound	$H_2SO_4/\%$	Temp. range/°C	$E/kJ\,mol^{-1}$	$\log_{10}(A/l\,mol^{-1}\,s^{-1})$	Ref.
	(< 89 %)				
Acetophenone	81·4	25–45	74	11·2	8
	88·8	25–45	69	12·1	8
Benzoic acid	81·4	25–45	58	9·4	8
2,6-Dichloropyridine*	81·5	75–104	113	11·1	13
2,6-Dichloropyridine 1-oxide	82·4	59–94	113	13·5	23
	87·9	64–104	107	14·2	23
2,6-Lutidine 1-oxide	78·2	65–95	111	11·8	10
	81·4	65–95	100	10·7	10
	84·3	65–95	97	10·8	10
	87·7	65–95	96	10·8	10
p-Nitroaniline	84·7	25–40	94	13·8	11
Pyridine 1-oxide†	87·9	95–125	98	9·8	10
Quinoline 1-oxide†	82·0	0–25	93	12·5	2
	(> 89 %)				
Acetophenone	98·1	25–45	75	11·3	8
2-Chloro-4-nitroaniline	93·8	25–40	82	13·6	11
	98·0	25–40	80	12·3	11
2,6-Dichloropyridine 1-oxide	94·9	64–94	98	13·8	23
3,5-Dichloropyridine 1-oxide	90·0	73–104	117	14·1	23
2,6-Dimethoxy-3-nitropyridine	89·9	15–37	82	15·9	13
3,5-Dimethoxy-2-nitropyridine	89·75	40–60	95	17·5	24
3,5-Dimethylpyridine 1-oxide	91·8	61–86	106	19·0	23
2,6-Lutidine 1-oxide	92·5	65–90	95	10·8	10
	97·8	65–90	103	11·2	10
3-Methyl-2-pyridone	94·2	22–37	59	·	9
5-Methyl-2-pyridone	94·2	29–38	80	·	9
p-Nitroaniline	90·1	25–40	80	12·2	11
	98·0	25–40	76	10·4	11

* The concentration of nitric acid was high: 81·5 % H_2SO_4, 11·6 % HNO_3, 5·9 % H_2O.
† For 4-nitration.

in this book, are exemplified by those mentioned in §5.3.4. Cases of the latter type have already been mentioned in this chapter. A striking example is that of quinoline 1-oxide; when this compound is nitrated at a particular temperature the proportion of 4-nitration decreases and that of $(5+8)$-nitration increases with increasing acidity. The 4-nitration involves the free base, and $(5+8)$-nitration involves the cation.[2] The nitration of cinnoline 2-oxide shows similar features as regards 6-substitution on the one hand and $(5+8)$-substitution on the other.[2]

The case of pyridine 1-oxide and its derivatives is also striking; with these compounds the orientation of nitration at $C_{(4)}$ stands out in contrast to the orientation of sulphonation and acid-catalysed deuteration at $C_{(3)}$,[22] and agrees with other evidence in suggesting that nitration involves the free bases.[10] Where the 4-position is already blocked, as in 2,6-dimethyl-4-methoxypyridine 1-oxide and related compounds, nitration of the 3-position occurs in the conjugate acids.[23] Orientation in the nitration of *p*-nitroaniline also agrees with other evidence in indicating nitration to occur in the free base.[11] These and related examples are discussed more generally in succeeding chapters in connection with substituent effects.

REFERENCES

1. Moodie, R. B., Schofield, K. & Williamson, M. J. (1964). *Nitro-Compounds*, Proceedings of International Symposium, Warsaw (1963), p. 89. London: Pergamon Press.
2. Gleghorn, J. T., Moodie, R. B., Qureshi, E. A. & Schofield, K. (1968). *J. chem. Soc. B*, p. 316.
3. Arnett, E. M. & Bushick, R. D. (1964). *J. Am. chem. Soc.* **86**, 1564.
4. Shapiro, S. A. (1969). Ph.D. thesis, University of East Anglia. Johnson, C. D., Katritzky, A. R. & Shapiro, S. A. (1969). *J. Am. chem. Soc.* **91**, 6654.
5. Deno, N. C. & Stein, R. (1956). *J. Am. chem. Soc.* **78**, 578.
6. Moodie, R. B., Qureshi, E. A., Schofield, K. & Gleghorn, J. T. (1968). *J. chem. Soc. B*, p. 312.
7. Austin, M. W., Blackborow, J. R., Ridd, J. H. & Smith, B. V. (1965). *J. chem. Soc.* p. 1051.
8. Moodie, R. B., Penton, J. R. & Schofield, K. (1969). *J. chem. Soc. B*, p. 578.
9. Brignell, B. J., Katritzky, A. R. & Tarhan, H. O. (1968). *J. chem. Soc. B*, p. 1477.
10. Gleghorn, J. T., Moodie, R. B., Schofield, K. & Williamson, M. J. (1966). *J. chem. Soc. B*, p. 870.
11. Hartshorn, S. R. & Ridd, J. H. (1968). *J. chem. Soc. B*, p. 1068.
12. Katritzky, A. R. & Ridgewell, B. J. (1963). *J. chem. Soc.* p. 3882.

References

13. Johnson, C. D., Katritzky, A. R., Ridgewell, B. J. & Viney, M. (1967). *J. chem. Soc.* B, p. 1204.
14. Perrin, D. D. (1964). *Aust. J. Chem.* **17**, 484.
15. Frost, A. A. & Pearson, R. G. (1953). *Kinetics and Mechanism.* New York: Wiley.
16. Modro, T. A. & Ridd, J. H. (1968). *J. chem. Soc.* B, p. 528.
17. Coombes, R. G., Crout, D. H. G., Hoggett, J. G., Moodie, R. B. & Schofield, K. (1970). *J. chem. Soc.* B, p. 347.
18. Brickman, M. & Ridd, J. H. (1965). *J. chem. Soc.* p. 6845.
19. Gillespie, R. J. & Norton, D. G. (1953). *J. chem. Soc.* p. 971.
20. Bonner, T. G., Bowyer, F. & Williams, G. (1952). *J. chem. Soc.* p. 3274.
21. Austin, M. W. & Ridd, J. H. (1963). *J. chem. Soc.* p. 4204.
22. Schofield, K. (1967). *Heteroaromatic Nitrogen Compounds: Pyrroles and Pyridines.* London: Butterworths.
23. Johnson, C. D., Katritzky, A. R., Shakir, N. & Viney, M. (1967). *J. chem. Soc.* B, p. 1213.
24. Johnson, C. D., Katritzky, A. R. & Viney, M. (1967). *J. chem. Soc.* B, p. 1211.

9 Nitration and aromatic reactivity: C. The nitration of monocyclic compounds

9.1 MONOSUBSTITUTED DERIVATIVES OF BENZENE

It is the purpose of this and the following chapter to report the quantitative data concerning the relationship of structure to orientation and reactivity in aromatic nitration. Where data obtained by modern analytical methods are available they are usually quoted in preference to the results of older work. Many of the papers containing the latter are, however, noted in the brief discussion which is given of interpretations of the results.

9.1.1 *Alkyl and substituted-alkyl groups*

Data for alkyl-benzenes are collected in table 9.1, and for substituted-alkyl compounds in table 9.2.

Although the partial rate factors for the alkylbenzenes vary somewhat with the experimental conditions, the main facts of the situation are perfectly clear:

(1) An alkyl group activates all nuclear positions, the o- and p-positions more than the m-position. The activation is not very strong.

(2) As the alkyl group changes from methyl to *tert*-butyl f_p increases. The $m:p$ ratio does not change much.

(3) Along the series f_o decreases, and so does the $o:p$-ratio.

The influence of alkyl groups has been attributed to the $+I$ effect operating primarily at the o- and p-positions (I), and somewhat less strongly at the m-position by relay. Alternatively, the effect is seen as stabilising the transition states for o- and p-substitution (II), more than

(I)   (II)   (III)

TABLE 9.1 The nitration of alkylbenzenes*

Compound	Temp/°C	Nitrating system	Relative rate	Isomer proportions (%)			Partial rate factors			Ref.
				ortho	meta	para	f_o	f_m	f_p	
Toluene	0	AcONO$_2$-Ac$_2$O	27	58.1	3.7	38.2	47	3.0	62	1
	30	AcONO$_2$-Ac$_2$O	23	58.4	4.4	37.2	40	3.0	51	1
	0	AcONO$_2$	27†	61.4	1.6	37.0	49.7	1.3	60.0	2
	25	AcONO$_2$	25.2	56.1±0.5	2.5±0.5	41.4±0.5	42.4	1.9	62.6	3
	25	AcONO$_2$-Ac$_2$O	23†	63.3±2.8	2.8±0.8	33.9±2.3	46.5	2.1	48.5	4
	30	HNO$_3$-MeNO$_2$	21	58.5	4.4	37.1	37	2.8	47	1
	25	HNO$_3$-MeNO$_2$	21†	61.7±3.0	1.9±0.5	36.4±3.0	38.9	1.3	45.8	4
	25	HNO$_3$-MeNO$_2$	26.4	61.5	3.1	35.4	49	2.5	56	5
	45	HNO$_3$-aq.AcOH	24	56.5	3.5	40.0	42	2.5	58	6
	25	HNO$_3$-AcOH	28.8	56.9	2.8	40.3	49	2.4	70	5
	25	HNO$_3$-sulpholan	17	61.9	3.5	34.7	32	1.7	35	5
	25	HNO$_3$-CF$_3$.CO$_2$H	28	61.6	2.6	35.8	51.7	2.18	60.1	7
	25	30% mixed acid in sulpholan	28	62.0	3.4	34.6	52.1	2.8	58.1	5
Ethylbenzene	0	AcONO$_2$	22.8±1.9	45.9	3.3	50.8	31.4	2.3	69.5	2
	25	HNO$_3$-MeNO$_2$	22.6	48.3	2.3	49.5	32.7	1.6	67.1	5
	25	30% mixed acid in sulpholan	24	50.3	3.6	46.1	36.2	2.6	66.4	5
iso-Propylbenzene	0	AcONO$_2$	17.7±0.7	28.0	4.5	67.5	14.8	2.4	71.6	2
	25	30% mixed acid in sulpholan	13.8	43.2	4.5	52.3	17.9	1.9	43.3	5
tert-Butylbenzene	45	HNO$_3$-aq.AcOH	15.6	12.0	8.5	79.5	5.5	4.0	75	6
	0	AcONO$_2$	15.1±0.8	10.0	6.8	83.2	4.5	3.0	75.5	2
	25	AcONO$_2$	14.9		.					3
	25	AcONO$_2$-Ac$_2$O	2.0±0.3‡	10.3±1.3	10.3±1.8	79.4±1.8	3.8	3.8	57.7	4
	25	HNO$_3$-MeNO$_2$	14±0.1‡	12.2±2.0	8.2±0.8	79.6±3.0	5.5	3.7	71.6	4

* See also tables 4.1 and 4.2   † Adopted from ref. 1.
‡ Ratio toluene-tert-butylbenzene. The partial rate factors are based on the relative rates for toluene:benzene of ref. 1.

it does that for *m*-substitution (III). The overall reactivity of the alkyl-benzenes decreases in the order $CH_3 > Et > i\text{-}Pr > t\text{-}Bu$, which might be seen as a consequence of the operation of hyperconjugation. However, the sequence is the result of the decrease in f_o along the series. That decrease is convincingly attributed to a primary steric effect, for the overall polar effects of the substituents do not show much change;[6, 23] this explanation was first suggested by Le Fèvre (§9.2.1).

When the *p*-positions are considered it is seen that they follow the sequence of inductive effects, and not of hyperconjugation. In this respect nitration is unusual amongst electrophilic substitutions.[24]

In a M.O. treatment of the electronic effect of the methyl group it was found necessary to take into account both inductive and hypercon-jugative effects.[25] This treatment is commented on in §9.3 below.

Consideration of the effects upon orienting and activating properties of the methyl group caused by replacing its hydrogen atoms with other groups throws light upon the way in which alkyl groups release electrons. The data are given in table 9.2 and the essential features for mono-substituted methyl groups are summarised below. From the first four

PhCH$_2$X*	H	Me	OMe	CO$_2$Et	Cl	CN	SO$_2$Et	NO$_2$
Relative rate	25	23	6·5	3·9	0·71	0·35	0·23	0·12
ortho	a	a	a	a	d	d	d	d
meta	a	a	a	a	d	d	d	d
para	a	a	a	a	a	a	d	d
Dominant orientation	o:p	o:p	o:p	o:p	o:p	o:p	o:p	m

a = activated; d = deactivated.   * For X = Ph see §10.1

examples we learn nothing which adds to the description of the methyl group as exerting a $+I$ effect; dominant o:p-orientation is associated with activation at all positions, although the degree of activation de-creases along the series. With the next two substituents overall deactiva-tion is found, but activation persists at the *p*-positions; o:p-orientation is still dominant. With the last two substituents there is general deacti-vation, but in one case this is accompanied by o:p- and in the other by *m*-direction. A $-I$ effect cannot alone account for such a situation; it would deactivate the o- and *p*-positions and, less markedly, the *m*-position also by relay. The association of deactivation with o:p-direction (an 'anti-Holleman' circumstance; §1.3) is thus seen as a manifestation of

TABLE 9.2   *The nitration of substituted-alkylbenzenes*

Compound	Temp./°C	Nitrating system	Relative rate	Isomer proportions/%			Partial rate factors			Ref.
				ortho	*meta*	*para*	f_o	f_m	f_p	
Benzyl methyl ether	25	$AcONO_2$	6·48	51·3	6·8	41·9	9·97	1·32	16·3	3
Ethyl phenylacetate	25	$AcONO_2$	3·86	54·3	13·1	32·6	6·29	1·52	7·55	3
Benzyl chloride	25	$AcONO_2$	0·711	33·6	13·9	52·5	0·716	0·296	2·24	3
Benzyl cyanide	25	$AcONO_2$	0·345	24·4	20·1	55·5	0·252	0·208	1·15	3
Ethyl benzyl sulphone*	25	HNO_3-Ac_2O	0·229	35·6	21·9	42·5	0·245	0·150	0·584	20a
ω-Nitrotoluene	25	HNO_3-Ac_2O	0·122	22·5	54·7	22·8	0·082	0·200	0·167	3
Benzyl fluoride	~25	$AcONO_2$-Ac_2O	.	28·3	17·3	54·4	.	.	.	8
ω-Toluenesulphonate	-10 to -5	HNO_3	.	33·5	13·7	52·8	.	.	.	9
Methyl benzyl sulphone	-10 to -5	HNO_3	.	27·5	30·3	42·2	.	.	.	9
ω-Toluenesulphon-amide	-10 to -5	HNO_3	.	27·9	31·4	40·7	.	.	.	9
Methyl ω-toluenesulphonate	-10 to -5	HNO_3	.	24·7	32·4	42·9	.	.	.	9
ω-Toluenesulphonyl chloride	-10 to -5	HNO_3	.	16·9	50·8	33·1	.	.	.	9
Benzal chloride	20-30	?	.	23·3	33·8	42·9	.	.	.	10
Benzotrichloride	20-30	?	.	6·8	64·5	28·7	.	.	.	10
Benzotrifluoride	0	HNO_3-H_2SO_4	.	6	91	3	.	.	.	11
	25	HNO_3-80·9%H_2SO_4	$2·6 \times 10^{-5}$	.	.	.	.	.	$6·7 \times 10^{-5}$†	12

* Nitric acid (d 1·529; 0·1 mol) was added to the sulphone (0·025 mol) in acetic anhydride (0·02 mol) at -70 °C and the temperature was allowed to rise to 25 °C.

† Using the proportion of *m*-nitration reported in ref. 11.

polarisability arising in hyperconjugative release from the methylene groups; this overcomes the inductive deactivation, except with the group $.CH_2NO_2$.[3, 20, 26]

Data for the other compounds in table 9.2 are less complete. The trihalogenomethyl groups are usually regarded as exerting powerful $-I$ effects, but the hyperconjugative properties of $.CF_3$ have been considered.[27]

9.1.2 Positive poles

Since the original observations of Vorländer, it has been recognised that positively charged substituents directly attached to the benzene ring are dominantly *m*-directing. Vorländer[28] examined the nitration of $Ph\overset{+}{N}Me_3$, Ph_3Bi^{2+}, Ph_2Pb^{2+}, Ph_2I^+ and 1-phenylpyridinium. Since Vorländer's time other examples have been studied and some results are summarized below. These early results for cations of elements of

Cation	*m*-Isomer %	Ref.	Cation	*m*-Isomer %	Ref.
PhHg+	50	29	PhNMe₃+	100	30
Ph₂Tl+	75	29	PhPMe₃+	100	30
PhTl²+	86	29	PhAsMe₃+	98	30
Ph₂Sn²+	79	29	PhSbMe₃+	86	30
Ph₂Pb²+	94	29	PhSMe₂+	100	22a
Ph₃Bi²+	86	29	PhSeMe₂+	100	22a
			Ph₂I+	82·5	29,31

group V indicated exclusive *m*-substitution in the first two cases. Combined with the results for the sulphur and selenium compounds, and for the related benzyl trimethyl derivatives of nitrogen, phosphorus, and arsenic,[21, 30, 32] and of sulphur and selenium,[22a] they established the general pattern that *m*-orientation increases as we cross a period, and decreases as we ascend a group of the periodic system.[33] The effect of a positive pole directly attached to the benzene ring was therefore seen as a consequence of a very powerful $-I$ effect generated by the positive charge (and relayed by conjugation in the ring; the π-inductive effect), and modified by the electronegativity of the particular central atom involved. Another view of the matter has already been mentioned (§7.2.1).

The way in which methylene groups interposed between the positive pole and the benzene ring, in cations of the type $Ph(CH_2)_n\overset{+}{X}$, weakened the *m*-orienting power of the pole received early attention,[21, 32, 34] and, as we shall see, this attention has been renewed more recently.

That some modification to the position so far described might be necessary was indicated by some experiments of Nesmeyanov and his co-workers.[35] Amongst other compounds they nitrated phenyl trimethyl ammonium and triphenyloxonium tetrafluoroborates; with mixed acid the former gave 96 % of *m*- and 4 % of *p*-nitro compound (88 % total yield), whilst the latter gave 80 % of the tri-(*p*-nitrophenyl)oxonium salt. Ridd and his co-workers have made a quantitative study of the phenyl trimethyl ammonium ion. Their results, and those of other recent workers on the nitration of several cations, are collected in table 9.3.

Included in table 9.3 are data for the cations $PhNH_3^+$, $PhNH_2Me^+$, $PhNHMe_2^+$, $PhCH_2NH_3^+$ and p-$Cl.C_6H_4.NH_3^+$. For each of these cations it has first to be established whether it or its conjugate base is being nitrated. Application of criteria discussed in §8.2 established that over the range 82–98 % sulphuric acid the anilinium ion is the entity being nitrated[14, 16a] and not the aniline.* The same is true over the more limited ranges of acidity studied (table 9.3) for $PhNH_2Me^+$ and $PhNHMe_2^+$,[15] and for p-$Cl.C_6H_4.NH_3^+$.[14] The case for the benzyl-ammonium ion is based mainly on analogy, but is not in doubt. It will be recalled that 2-chloro-4-nitroaniline and *p*-nitroaniline show a different kind of behaviour; at high acidities the free bases, present in the solutions in small concentrations, are nitrated upon encounter (§8.2).

We shall now consider the implications of these newer results for the nitration of these cations, taking first the comparison of the anilinium ion with its increasingly methylated homologues, then the various cations containing the trimethylammonio group, and finally cations containing elements other than nitrogen.

As regards the anilinium ion[13-15] (table 9.3) we see (data for nitration in 98 % sulphuric acid) that the *p*-position is more reactive than a *m*-position, a result which would not be expected on the basis of the $-I$ effect transmitted conjugatively in the ring. In the series $PhNMe_3^+$, $PhNHMe_2^+$, $PhNH_2Me^+$, $PhNH_3^+$ there is seen a smooth change from predominant *m*- to almost equal *m*- and *p*-reactivity, though the differences are small when compared with the powerful, overall de-

---

* At lower acidities the observed rate constant for nitration of *p*-nitroaniline becomes greater than that for nitration of anilinium.[16]

activating influences of these substituents. The change from .NH_3^+ to .NMe_3^+ decreases the rate of *p*- even more than the rate of *m*-nitration, a marked contrast to the effect of going from methyl to *tert*-butyl in the isoelectronic alkyl series.

The preferred explanation of these results is that in the protonated cations hydrogen bonding ameliorates the direct field effect by spreading the positive charge. This does not account for all of the facts; thus, whilst in the sequence $PhNMe_3^+$, $PhNH_3^+$, $PhCH_2NMe_3^+$ the relative rate of nitration increases, the proportion of *p*-nitration varies differently (11 %, 38 % and 15 % in 98 % sulphuric acid). The anilinium ion is intermediate in reactivity but gives more *p*-isomer than either of the other two. It seems that both orientation and reactivity cannot be described by varying a single parameter controlling the interaction between the pole and the benzene ring. That the protonated poles give higher *p*:*m*-ratios than are expected from relative reactivities may be due to the way in which the positive charge is spread in the medium, or to hyperconjugative release of electrons from the protonated poles.

The combined inductive and field effects of these poles do not produce strong discrimination between the *m*- and *p*-positions in nitration ($\frac{1}{2}$ *m*:*p* ~ 4 for .NMe_3^+, and smaller for the protonated poles). This situation is in marked contrast to that produced by, say, the nitro group (§9.1.3), and suggests that the $-M$ effect is more discriminating between *m*- and *p*-positions than is the $-I$ effect.

As has been noted above, there is no gross change in the mechanism of nitration of $PhNH_3^+$ down to 82 % sulphuric acid. The increase in *o*- and *p*-substitution at lower acidities has been attributed[16a] to differential salt effects upon nitration at the individual positions. The two sets of partial rate factors* quoted for $PhNH_3^+$ in table 9.3 show the effect of the substituent on the Gibbs function of activation at the *m*- and *p*-positions to be roughly equal for reaction in 98 % sulphuric acid, and about 28 % greater at the *p*-position in 82 % sulphuric acid.[16a]

These results reveal the positive poles as having rather different characters from those previously attributed to them; according to the older view they were very strongly *m*-directing, a characteristic which is now seen to be much weaker than was thought. Lack of knowledge of partial rate factors led to earlier overestimates of the effect.[36] Further consideration of the effects of these substituents by examining the way in which they influence the Gibbs functions of activation at *m*- and

* Concerning the validity of the partial rate factors see ref. 15. See also §7.1.2.

TABLE 9.3 The nitration of benzene derivatives containing positively charged substituents*

Cation	Nitrating system†	Isomer proportions %			Rate constants‡/l mol⁻¹ s⁻¹				Partial rate factors			Ref.
		ortho	meta	para	k_2	k_2^o	k_2^m	k_2^p	f_o	f_m	f_p	
$PhNH_3^+$	82·0 % H₂SO₄	5	36	59	0·012	0·31$\times10^{-2}$	0·22$\times10^{-2}$	0·74$\times10^{-2}$	19$\times10^{-8}$	138$\times10^{-8}$	451$\times10^{-8}$	16a
	84·0 % H₂SO₄	5	36	59	0·079	0·20$\times10^{-2}$	1·4$\times10^{-2}$	4·7$\times10^{-2}$				16a
	84·9 % H₂SO₄	4	37	59	0·16	0·32$\times10^{-2}$	3·0$\times10^{-2}$	9·7$\times10^{-2}$				16a
	86·5 % H₂SO₄	4	39	57	0·66	1·3$\times10^{-2}$	13$\times10^{-2}$	38$\times10^{-2}$				16a
	87·5 % H₂SO₄	3	42	55	1·1	1·6$\times10^{-2}$	23$\times10^{-2}$	60$\times10^{-2}$				16a
	88·5 % H₂SO₄	3	45	52	2·0	3·0$\times10^{-2}$	45·5$\times10^{-2}$	100$\times10^{-2}$				16a
	89·5 % H₂SO₄	3	48	49	2·6	3·9$\times10^{-2}$	63$\times10^{-2}$	130$\times10^{-2}$				16a
	92·4 % H₂SO₄	≯2	53	47	2·01		0·533	0·945				13–15
	94·8 % H₂SO₄	≯2	57	43	1·47		0·419	0·632				13–15
	96·4 % H₂SO₄	≯2	58	42	1·08		0·313	0·454				13–15
	98·0 % H₂SO₄	1·5	62	38	0·668	0·005	0·207	0·254	4·3$\times10^{-8}$	173$\times10^{-8}$	213$\times10^{-8}$	13–16
	100 % H₂SO₄	≯2	64	36·5	0·655		0·210	0·236				13–15
$PhNH_2Me^+$	90·9 % H₂SO₄		61	39	0·747		0·228	0·291				15
	96·2 % H₂SO₄		67·5	32·5	0·317		0·107	0·103				15
	98·0 % H₂SO₄		(70)	30			0·074	0·064				13, 15
	99·8 % H₂SO₄		70·3	29·7	0·127		4·46$\times10^{-2}$	3·77$\times10^{-2}$		57$\times10^{-8}$	49$\times10^{-8}$	15
$PhNHMe_2^+$	91·1 % H₂SO₄		74	26	0·135		4·99$\times10^{-2}$	3·51$\times10^{-2}$				15
	96·2 % H₂SO₄		76	24	5·92$\times10^{-2}$		2·25$\times10^{-2}$	1·42$\times10^{-2}$				15
	98·0 % H₂SO₄		78	22			1·6$\times10^{-2}$	0·93$\times10^{-2}$				13, 15
	100 % H₂SO₄		79	21	2·68$\times10^{-2}$		1·06$\times10^{-2}$	0·56$\times10^{-2}$		12·3$\times10^{-8}$	7·1$\times10^{-8}$	15
$PhNMe_3^+$	98·0 % H₂SO₄		89	11	1·04$\times10^{-2}$		5·5$\times10^{-3}$	1·4$\times10^{-3}$		4·67$\times10^{-8}$	1·15$\times10^{-8}$	13, 15
	98·7 % H₂SO₄	≯2	89	11			4·6$\times10^{-2}$	1·1$\times10^{-3}$				17
$PhCH_2NH_3^+$	78·7 % H₂SO₄				1·58							13, 15
	80·0 % H₂SO₄				3·72							13, 15
$PhCH_2NMe_3^+$	74·5 % H₂SO₄				0·056$\times10^{-2}$							18
	76·5 % H₂SO₄				0·293$\times10^{-2}$							18
	78·7 % H₂SO₄				0·252							13, 15
	79·3 % H₂SO₄				0·345							18
	80·0 % H₂SO₄		85§	15§	0·600							13, 15
	81·6 % H₂SO₄				0·283							18

Cation	Nitrating system†	Isomer proportions %			Rate constants‡ /l mol⁻¹ s⁻¹				Partial rate factors			Ref.
		ortho	meta	para	k_2	k_2^o	k_2^m	k_2^p	f_o	f_m	f_p	
$Ph(CH_2)_2NMe_3^+$	63.4 % H_2SO_4	.	19§	.	0.0195×10^{-2}	.	.	.	.	.	.	18
	68.3 % H_2SO_4	.	.	.	0.823×10^{-2}	.	.	.	.	.	.	18
	72.7 % H_2SO_4	.	.	.	35.7×10^{-2}	.	.	.	.	.	.	18
	76.5 % H_2SO_4	.	.	.	80.4×10^{-2}	.	.	.	.	.	.	18
$Ph(CH_2)_3NMe_3^+$	61.0 % H_2SO_4	.	5 ± 2§	.	0.0482×10^{-2}	.	.	.	.	.	.	18
	63.4 % H_2SO_4	.	.	.	0.231×10^{-2}	.	.	.	.	.	.	18
	68.3 % H_2SO_4	.	.	.	10.5×10^{-2}	.	.	.	.	.	.	18
$p\text{-Cl}.C_6H_4NH_3^+$	91.0 % H_2SO_4	.	.	.	0.231	.	.	.	.	.	.	14
	99.5 % H_2SO_4	.	.	.	0.056	.	.	.	.	.	.	14
$p\text{-Cl}.C_6H_4NMe_3^+$	91.0 % H_2SO_4	.	.	.	0.0058	.	.	.	.	.	.	14
	99.5 % H_2SO_4	.	.	.	0.0015	.	.	.	.	.	.	14
$PhPMe_3^+$	98.7 % H_2SO_4	.	98‖	~2‖	0.0501	.	.	.	.	24.7×10^{-8}	~1.0×10^{-8}	17
$PhCH_2PMe_3^+$	HNO_3-$MeNO_2$	13.1	19.4	67.5	.	.	.	.	0.0026	0.0039	0.0268	20b
$PhAsMe_3^+$	98.7 % H_2SO_4	o + p = 4‖	.	m = 96‖	0.400	.	.	.	.	194×10^{-8}	<16.2×10^{-8}	17
$PhCH_2AsMe_3^+$	HNO_3-$MeNO_2$	17.5	6.7	75.7	.	.	.	.	0.0067	0.0026	0.0577	20b
$PhSbMe_3^+$	75.9 % H_2SO_4	8‖	87‖	5‖	0.115×10^{-2}	.	.	.	840×10^{-8}	9140×10^{-8}	1050×10^{-8}	17
	76.5 % H_2SO_4	.	.	.	0.183×10^{-2}	.	.	.				
	79.3 % H_2SO_4	.	.	.	2.58×10^{-2}	.	.	.				
	80.7 % H_2SO_4	.	.	.	7.65×10^{-2}	.	.	.				
	81.6 % H_2SO_4	.	.	.	19.2×10^{-2}	.	.	.				
	82.1 % H_2SO_4	.	.	.	36×10^{-2}	.	.	.				
$PhSMe_2^+$	Mixed acid	3.6 ± 0.2	90.4 ± 0.3	6.0 ± 0.2	.	.	.	.	.	.	.	19
$PhCH_2SMe_2^+$	HNO_3(−15 °C)	.	52	.	.	.	.	.	.	.	.	22a
$PhSeMe_2^+$	Mixed acid	2.6 ± 2	91.3 ± 0.3	6.1 ± 0.2	.	.	.	.	.	.	.	19
$PhCH_2SeMe_2^+$	HNO_3(−15 °C)	.	16	.	.	.	.	.	.	.	.	22a

* At 25 °C.

† Solutions of nitric acid (d 1.42) in sulphuric acid of the given concentration.

‡ Rate = k_2 [substrate] [HNO_3].

§ These isomer proportions were obtained for nitration with nitric acid at 0° to −15 °C. The figure reported for $PhCH_2NMe_3^+$ was actually 88 % of the meta-isomer.[21]

‖ For the conditions under which the isomer proportions were determined the original paper should be consulted.[17]

p-positions throws light on their mode of operations.[37] The partial rate factor, f, permits the calculation of the change in Gibbs' function of activation effected by a substituent at the position to which it refers ($\delta\Delta G^{\ddagger} = -RT \ln f$); in particular the new data permit the calculation of $\delta\Delta G_p^{\ddagger}/\delta\Delta G_m^{\ddagger}$, and some values for nitration follow. As regards the positive poles, the difference between the effect of each at p- and m-positions is very small, whilst the corresponding effect of an alkyl group at the p-position is by comparison much larger at the p- than at the m-position. This contrast is taken to indicate that the poles and the alkyl groups, both of which are regarded as inductive substituents, exert substituent effects which differ in their modes of operation as well as in their directions. As regards the alkyl groups the situation is what would be expected from the discussion already given (§9.1.1); the π-inductive effect, perhaps aided by hyperconjugation, produces activation of the p- as against the m-position.

Substituent	$\delta\Delta G_p^{\ddagger}/\delta\Delta G_m^{\ddagger}$	Substituent	$\delta\Delta G_p^{\ddagger}/\delta\Delta G_m^{\ddagger}$
$\cdot NMe_3{}^+$	1·09	$\cdot CMe_3$	3·12
$\cdot NH_3{}^+$	0·99	$\cdot CH_3$	4·43

With the cations the closely similar reactivities of p- and m-positions reveal a substituent effect which causes deactivation of the ring without much discrimination between p- and m-positions; such a substituent effect is seen as arising from the field effect, which on a simple picture of the transition state, and depending on the distribution of the charge in the transition state, could slightly favour either position.[37] Support for this view is seen in the fact that the deactivating power of a positive pole falls off far less rapidly with distance from the ring than is the case with a neutral substituent (the case of $PhCH_2NO_2$ and $PhNO_2$ (see below) should be compared with a corresponding pair of cations). Also, for a given degree of overall deactivation the poles produce more p-substitution than do neutral substituents; nitrobenzene and phenyl trimethyl ammonium are of similar gross reactivity, but the latter produces considerably more of the p-isomer in nitration than does the former.

Further light on the substituent effects of nitrogen poles comes from the data for the family of compounds $Ph(CH_2)_n \cdot \overset{+}{N}Me_3$ ($n = $ 0–3).[18] Over the whole series the relative reactivities of these ions with respect to benzene ($3\cdot16$, $0\cdot224$, $7\cdot94 \times 10^{-5}$, and $3\cdot39 \times 10^{-8}$ in decreasing order of chain length) are not a satisfactory measure of the effect of the pole because of the intervention of the activating methylene groups. To remove the latter effect interest was centred on m-positions and the quantity $-\log f_m/f_m^\circ$ (where f_m° is the partial rate factor for the m-position in the related alkylbenzene), which is proportional to the increase in Gibbs' function of activation caused by replacing a terminal hydrogen atom of an alkyl chain by the trimethylammonio group, was evaluated. The way in which this quantity changed with chain length was not in accord with the view that the poles act inductively through the methylene chain, but resembled the way in which a methylene group affected the dissociation constants of dicarboxylic acids (§7.2.1). The point was illustrated by plotting $-\log f_m/f_m^\circ$ against r_x calculated from the simple model for the transition state illustrated below, and the corresponding

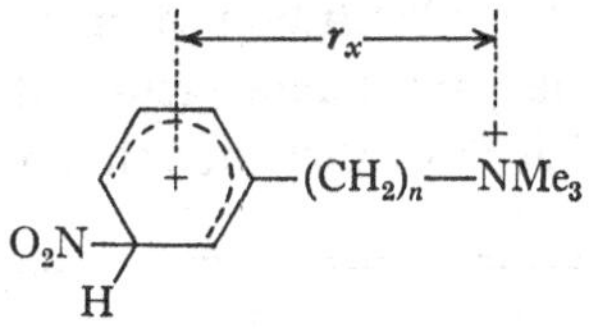

pair of quantities for the ionisation of the anions $HO_2C.(CH_2)_n.CO_2^-$; both sets of data fell on one straight line. The substituent effects of both positive and negative poles are evidently, in the main, consequences of direct electrostatic interactions. The effect of varying the chain length in the cations upon the activation parameters for nitration supported this view.

If this electrostatic treatment of the substituent effect of poles is sound, the effect of a pole upon the Gibbs' function of activation at a particular position should be inversely proportional to the effective dielectric constant, and the longer the methylene chain the more closely should the effective dielectric constant approach the dielectric constant of the medium. Surprisingly, competitive nitrations of phenpropyl trimethyl ammonium perchlorate and benzene in acetic anhydride and tri-fluoroacetic acid showed the relative rate not to decrease markedly with the dielectric constant of the solvent. It was suggested that the expected decrease in reactivity of the cation was obscured by the faster nitration of ion pairs.

The general conclusion drawn from this work was that to a first approximation the electrostatic interaction between the nitrogen poles and the charge on the ring in the transition state gave the best account of the substituent effects of the poles.[18]

It will be noticed that this account makes no allowance for the electrostatic interaction of the positive pole with the electrophile, the nitronium ion. This should generally work for deactivation, and its influence at nuclear positions should be in the order *ortho* > *meta* > *para*. This point is resumed below.

As already noticed, it was early appreciated that the effect of a positive pole on substitution would depend on the electronegativity of the atom concerned, as well as upon the fact that it carried a unit positive charge. The important facts (table 9.3) are that for the cations $PhXMe_3^+$ reactivity increases with variation of X in the order $N < P < As < Sb$, that the amount of p-substitution is greater when $X = N$ than when $X = Sb$ and goes through a minimum with the other two elements, and that the amount of o-substitution is greater when $X = Sb$ than when $X = N$. From the partial rate factors the following quantities can be calculated,[17] which show that the other atoms produce a greater

	NMe_3^+	PMe_3^+	$AsMe_3^+$	$SbMe_3^+$
$\delta\Delta G_p^\ddagger / \delta\Delta G_m^\ddagger$	1·08	1·21	1·19	1·23

increase in $\delta\Delta G_p^\ddagger$ as compared with $\delta\Delta G_m^\ddagger$ than does nitrogen. The differences arise because f_m increases with the atomic number of X whilst f_p does not. It seems that some effect operating mainly at the p-position is present which either helps p-substitution in the phenyl trimethyl ammonium ion or retards it in the other cases. The second alternative is preferred, and is identified as a $-M$ effect in the phosphonium, arsonium, and stibonium poles (which are therefore $-I$ $-M$ groups), arising from overlap of the aromatic π-orbitals with vacant d-orbitals in these cases. As regards the results for o-substitution, it may be recalled that $-M$ effects operate much more strongly at p- than at o-positions (cf. the nitro-group; §9.1.3).[17]

The problem of electrophilic substitution into the anilinium ion has been examined by the methods of M.O. theory. Attempts to simulate the π-inductive effect in Hückel M.O. theory by varying the Coulomb integral of $C_{(1)}$ (the carbon atom to which the $\cdot NH_3^+$ group is attached) remove π-electrons from the o- and p-positions and add them to the *m*-position. This model, as the basis of an isolated molecule theory, or of a transition state theory which assumed that bonding with the electrophile had not proceeded very far in the transition state, would thus predict almost exclusive *m*-nitration, which is not found. This description of the anilinium ion was modified in three stages.[38] First, the electrostatic effect of the positive pole was allowed to modify the Coulomb term of each atom (*r*) of the ring by an amount $1/\epsilon R_r$; that is, a field effect influencing all of the atoms, and not just $C_{(1)}$, according to their distance from the positive charge, was introduced, Secondly, the Hückel approximation was replaced by the Pople perturbation treatment. The consequence of these two modifications was a ground state of the anilinium cation in which all three positions carried positive charges in the order *para* > *ortho* > *meta*; clearly an isolated molecule theory based on such treatments could not correctly reproduce the substituent effect of the ammonio group. In the third stage of elaboration the second type of perturbation, caused by the approach of the nitronium ion, was introduced; charge distributions at the *o*-, *m*-, and *p*-positions were markedly changed, but again did not lead to the observed orientation of substitution. However, when with the state of affairs so reached the force on the nitronium ion approaching each position, or the total electrostatic energy was calculated, the *m*- and *p*-positions came out very similarly and both were markedly favoured over the *o*-position as points of attack. The other theoretical treatment was more elaborate, but similar in its conclusions.[39]

Earlier in this book (§7.2.1) the direct field effect was mentioned in rather general terms. The foregoing discussion brings out the fact that there are three aspects of the effect which have to be considered:

(1) The electrostatic interaction of the charge on the orienting substituent and the charge on the ring, or the ring positions in the transition state.

(2) The electrostatic interaction of the charge on the orienting substituent and the ring positions in the ground state.

(3) The electrostatic interaction of the charge on the orienting substituent, and those at the nuclear positions, with that of the approaching electrophile.

Consideration of (1), as in the work of Ridd and his co-workers, would constitute a transition state theory of the substituent effects. (2) alone would give an isolated molecule description, and (3), in so far as the charge on the electrophile was considered to modify those on the nuclear positions, would be an intermediate kind of description, or one which, alternatively, considered the reaction as one for which the transition state was formed early. Both of the elements (2) and (3) appear in the M.O. treatment described above, (2) alone being judged inadequate.

The interesting case of the triphenyloxonium ion[35] has already been mentioned. It has not been studied quantitatively, but the high degree of p-substitution reported for it would appear to make the oxonium substituent the only known case of a $-I + M$ group among poles. The oxygen atom presumably uses a lone-pair of electrons in conjugation with the ring. The case stands in marked contrast to that of the triphenylsulphonium ion, which is m-nitrated.[35]

9.1.3 *Dipolar groups*

The data which can be regarded as having some degree of quantitative significance, for the nitration of benzene derivatives containing substituents of this kind, are collected in table 9.4. Two features of the contents of this table need to be noticed before substituent effects in this group can be discussed: first, the data are very incomplete, and somewhat ill-assorted as a basis for comparisons; secondly, they refer to compounds which are in the main either weak or very weak bases. The usual criteria (§8.2) show that the nitrations of NN-dimethylaniline N-oxide and benzamide in 81–93 % sulphuric acid involve the conjugate acids of these compounds. Accordingly, if complete data on orientation become available for these compounds reacting under these conditions, they will be more appropriately discussed in §9.1.2 above. In contrast, benzoic acid and acetophenone, up to acidities at least as high as 90 % sulphuric acid, are nitrated as the free bases. For benzoic acid there is no evidence about the effect of acidity upon orientation, but that relating to acetophenone does seem to indicate an increase in m-substitution at the highest acidities. This increase has been ascribed to protonation,[22b] but is more likely to be due to a medium effect (hydrogen bonding) stopping short of proton transfer. The same may be true for benzaldehyde. The behaviour of the activity coefficients of nitro-compounds at high acidities has already been mentioned (§2.3.2), but it is not known if this influences orientation in the nitration of nitrobenzene. It is true that for nitrobenzene the data suggest a marked change in the $\frac{1}{2}$ m:p- and $\frac{1}{2}$ o:p-ratios as between nitration in nitric acid and nitration in sulphuric acid, but these are brought about by proportionately large changes in the absolutely small degree of p-substitution. Before the significance of the changes could be discussed the older data for isomer proportions might need to be checked.

If interest is centred upon the nitration of compounds of this type under comparable conditions (i.e. in nitric acid) which at the same time are less likely to produce the medium effects associated with nitration in the much stronger sulphuric acid, the feature which has attracted most attention becomes apparent; that is, the generally high $\frac{1}{2}o:p$-ratios. These are generally greater than unity, sometimes considerably greater. With such compounds as nitrobenzene, benzoic acid, benzonitrile, benzamide and ethyl benzoate the resulting situation is in marked contrast to those found with other compounds so far discussed and involving $-I$ effects, for which the $\frac{1}{2}o:p$-ratios are less than unity (table 9.2). It has been suggested that the high $\frac{1}{2}o:p$-ratios arise from a special mechanism existing for o-substitution[53] (cf. the cases discussed in §5.3.4), that these ratios are not unusual and that lower values in other cases arise from steric hindrance,[54] or that with these $-I -M$ substituents the p-position suffers special deactivation by the $-M$ effect.[55] The fact that with the $-I -M$ substituents the $\frac{1}{2}m:p$- and $\frac{1}{2}o:p$-ratios change together, decreasing towards unity among the compounds in table 9.4 (and continuing below unity for the compounds mentioned in §9.1.4 below) is seen as crucial in supporting the third explanation.[56] If the conjugative mechanism is always more effective at the p- than at the o-position, whether the overall result is determined by polarisation (highest values for both ratios) or by polarisability (decreasing values for both ratios), the pattern of table 9.4 would result. Results for other electrophilic substitutions support this description,[57] and the case of benzonitrile, in which the substituent is linear, is particularly difficult to fit to other explanations.

A different explanation of the high $o:p$-ratios is based on the view, for which there is some evidence, that in a transition state for substitution which resembles the Wheland intermediate in structure there is a larger positive charge at the p- than at the o-position. Substituents of the present type would therefore stabilise the transition state more from the o- than from the p-position.[57]

In some situations the nitro group behaves as if it exerted its influence mainly by the inductive process, but in nitration its behaviour seems to place it with this group of $-I -M$ substituents. The precise way in which a sulphone group is described depends on how much weight is given to the ability of the sulphur atom to expand its octet; the positive

TABLE 9.4 The nitration of benzene derivatives containing dipolar substituents

Compound	Nitrating system	Temp./°C	ortho	meta	para	k_2/l mol^{-1} s^{-1}	Relative rate	f_o	f_m	f_p	½ m:p	½ o:p	Ref.
Nitrobenzene*	HNO₃	0	6·4	93·2	<0·25						155	11	40
		30	8·1	91·2	0·7								40
	HNO₃-H₂SO₄	0	4·75	93·9	1·39						33·8	1·69	} 41
		25	6·12	91·8	2·06		$5\cdot8 \times 10^{-8}$	$1\cdot08 \times 10^{-8}$	$16\cdot2 \times 10^{-8}$	$0\cdot726 \times 10^{-8}$	22·3	1·48	
		40	6·74	90·9	2·35						19·3	1·43	
Benzoic acid†	HNO₃	−30	14·4	85·0	0·6						31	7	40
		30	18·5	80·2	1·3								44
		30	22·3	76·5	1·2								
		0	17	82	1								45
	HNO₃-81·4% H₂SO₄	25	·	·	·	$20\cdot3 \times 10^{-2}$ (k_2 obs.) $40\cdot5 \times 10^{-2}$ (k_2 fb.)	$39\cdot2 \times 10^{-2}$						45
Benzonitrile	HNO₃	0	17	81	<2								49
		−20	15·4	83·4	1·4								
		−9	16·1	82·1	1·6								
		0	16·8	80·8	1·95						20·7	4·3	50
Methyl phenyl sulphone	KNO₃-H₂SO₄	90	·	96–98	·								51
Ethyl phenyl sulphone	HNO₃-Ac₂O	25	8·1	88·6	3·3		$3\cdot51 \times 10^{-3}$	$0\cdot9 \times 10^{-3}$	$9\cdot3 \times 10^{-3}$	$0\cdot7 \times 10^{-3}$	13·4	1·2	20a
Benzamide†	HNO₃	15	27	69·6	<3						12	4·5	47
	HNO₃-81·2% H₂SO₄	25	·	·	·	$5\cdot68 \times 10^{-4}$	$7\cdot3 \times 10^{-3}$						45
Ethyl benzoate	HNO₃	−4	25·5	73·2	1·3								40
		0	28·3	68·4	3·3						10	4·3	40
		30	27·7	66·4	5·9								
	AcONO₂-Ac₂O	18	24·1	72·0	4·0		$3\cdot67 \times 10^{-3}$	$2\cdot6 \times 10^{-3}$	$7\cdot9 \times 10^{-3}$	$0\cdot9 \times 10^{-3}$	9	3	48

Compound	Nitrating system Temp./°C	Isomer proportions (%) ortho	meta	para	k_2/l mol⁻¹ s⁻¹	Relative rate	f_o	f_m	f_p	½ m:f	⅓ o:f	Ref.
Acetophenone† HNO₃·7 % oleum	−8 to 10	.	90·0	.	.	.	.	.	.	.	.	22b
HNO₃·80 % H₂SO₄	−8 to 10	.	83·1	.	.	.	.	.	.	.	.	
HNO₃ (d 1·505)		.	68·1	.	.	.	.	.	.	.	.	
HNO₃·80·3 % H₂SO₄	25	26·4	71·6	.	.	.	.	.	.	<17·9	<6·6	45
HNO₃·81·4 % H₂SO₄	25	.	.	0–2	2·08 × 10⁻³ [k_2 fb. = 15.2 × 10⁻²]	12·9 × 10⁻⁶	.	10·22 × 10⁻⁶	15·5 × 10⁻⁷			
HNO₃·85·1 %/H₂SO₄	25	26·2	71·8	0–2	19·6 × 10⁻² (k_2 obs.)	.	.	27·71 × 10⁻⁶	.	<17·9	<6·5	
HNO₃·98·1 % H₂SO₄	25	19·5	78·5	0–2	1·90 × 10⁻³	.	.	.	.	<19·6	<4·9	
Benzaldehyde HNO₃·7·3 % oleum	−8 to 10	.	90·8	.	.	.	.	.	.	.	.	22b
HNO₃·80 % H₂SO₄	−8 to 10	.	83·9	.	.	.	.	.	.	.	.	
HNO₃ (d 1·53)	−8 to 10	.	69·8	.	.	.	.	.	.	.	.	
HNO₃ (d 1·505)	−8 to 10	(19)	72·1	(9)	.	.	.	.	.	.	.	
Benzenesulphonic acid HNO₃–H₂SO₄	.	.	60	.	.	.	.	.	.	4	1·6	52
NN-Dimethyl-aniline N-oxide‖ HNO₃–84·4 % H₂SO₄	25	.	.	.	0·146 × 10⁻²	4·4 × 10⁻⁹	.	.	.	.	.	45

* The relative rate[41] is derived from the kinetic data[42a] by stepwise comparison with m-nitrotoluene, chlorobenzene and benzene. Kinetic data are available for the acidity range 80·0–95·6 % sulphuric acid[42]. See also ref. 43.

† The relative rate is calculated from the ratio[44] p-dichlorobenzene:benzene = 5·9 × 10⁻⁴. Kinetic data for the acidity range 78·0–81·4 % sulphuric acid and Arrhenius parameters for 81·4 % sulphuric acid are available.[45]

‡ The relative rate is again calculated from that of p-dichlorobenzene. Kinetic data for the acidity range 81·2–87·8 % sulphuric acid and Arrhenius parameters for 81·2 % sulphuric acid are available.[46]

§ Kinetic data are available for the range 75·5–98·1 % sulphuric acid, and Arrhenius parameters at several acidities.[45] The relative rate was obtained as before.[45]

‖ Kinetic data for the range 84·4–92·3 % sulphuric acid are available.[45] The relative rate was obtained as before.

charge on the sulphur atom has been regarded as important, particularly since benzenesulphonic acid, believed because it is a strong acid to be nitrated as its anion, still gives dominant *m*-substitution.

9.1.4 *Groups with 'lone pairs' conjugated to the ring*

The halogen substituents are the most important members of this group and the most important of the 'anti-Holleman' groups. The facts concerning their influences in nitration (table 9.5) are:

(1) They deactivate all positions, overall reactivities depending upon the substituent in the sequence F $\sim$ I > Cl $\sim$ Br.

(2) They deactivate *m*-positions most of all. Their influences in this connection do not differ greatly one from another but fluoro seems to be the most, and iodo the least deactivating.

(3) The $\frac{1}{2}$ *o*:*p*-ratio is always less than unity and varies with the substituent in the sequence I > Br > Cl > F.

(4) The $\frac{1}{2}$ *m*:*p*-ratio behaves like the $\frac{1}{2}$ *o*:*p*-ratio.

The halogen substituents $(-I +M)$ owe their *o*:*p*-orientating effect, achieved in spite of the deactivation, to polarisability by the conjugative process.[59] The strength of the inductive deactivation is seen in the sequence of the two ratios quoted.

In considering the sequence of overall reactivities, it is postulated that the inductive effects $(-I)$, F > Cl > Br > I, are modified by the electromeric polarisabilities, in the order I > Br > Cl > F, leading to the observed 'parabolic' sequence, F > Cl $\sim$ Br < I. There are difficulties which face this argument, in particular, that electrophiles which would be expected to be weakly polarising, such as molecular chlorine, seem to elicit a larger response from the fluoro substituent than does the more strongly polarising nitronium ion.[60] This difficulty is met by supposing that in the transition state for chlorination, bond-formation has proceeded further than it has in nitration.

We have seen already (§§9.1.2, 7.2.4) that attempts have been made to account for substituent effects, particularly those of the halogens, in electrostatic terms. In a simple M.O. treatment, using the Hückel approximation, variation of the halogen atom from fluorine through to iodine is simulated by decreasing the Coulomb integral for the hetero-atom, and also by decreasing the resonance integral of the carbon–halogen bond, i.e. by decreasing both h_x and k_{c-x} in the usual notation. Decreasing h_x alone increases q_r at the *o*- and *p*-positions, whilst decreasing k_{c-x} has the opposite effect. The overall result is a 'parabolic'

TABLE 9.5 *The nitration of the halogenobenzenes*

Compound	Nitrating system	Temp/°C	Isomer proportions/%*			k_2/l mol^{-1} s^{-1} †	Relative rate	f_o	f_m	f_p	½ o:p- ratio	½ m:p- ratio	Ref.
			ortho	meta	para								
Fluorobenzene	AcONO₂	25	8·7	0	91·3	·	0·141	0·04	0	0·77	0·047	·	2
	HNO₃-67·5 % H₂SO₄	25	13	0·6	86	3·9 × 10⁻³	0·117	0·045	0·0021	0·58	0·075	0·0034	46
	HNO₃	0	12·6	·	87·4	·	0·15‡	0·041	·	0·79	0·072	·	40
Chlorobenzene	AcONO₂-MeNO₂	25	29·6	0·9	69·5	·	0·033‡	0·029	0·0009	0·137	0·21	0·0064	58
	HNO₃-67·5 % H₂SO₄	25	35	0·94	64	1·8 × 10⁻³	0·064	0·067	0·0018	0·246	0·27	0·0073	46
Bromobenzene	AcONO₂-MeNO₂	25	36·5	1·2	62·4	·	0·030‡	0·033	0·0011	0·112	0·29	0·0096	58
	HNO₃-67·5 % H₂SO₄	25	43	0·9	56	1·4 × 10⁻³	0·060	0·077	0·0016	0·200	0·38	0·0080	46
Iodobenzene	AcONO₂-MeNO₂	25	38·3	1·8	59·7	·	0·22	0·252	0·012	0·78	0·32	0·015	58
	HNO₃-67·5 % H₂SO₄	25	45	1·3	54	6·2 × 10⁻³	0·125	0·17	0·0049	0·400	0·41	0·012	46

* The results are only slightly dependent upon conditions; for nitrations with acetyl nitrate see ref. 59, and in sulphuric acid ref. 46. The rate constants given refer to the acidities quoted, but the accompanying isomer proportions usually refer to slightly different acidities. However, as noted, isomer proportions are not much affected by changes in acidity. Rate profiles are available for all of the compounds.[46]

† For nitration with acetyl nitrate in acetic anhydride at 18 °C.

trend in q_r with high values for fluorine and iodine and lower ones for chlorine and bromine. As the basis of an isolated molecule theory this model would predict o:p-substitution with activation. However, pole–dipole interactions of the charge on the electrophile and the dipole of the carbon–halogen bond cause unfavourable repulsions varying from position to position in the order *ortho* ≫ *meta* > *para*. The combined result could then be o:p-substitution with deactivation, electrostatic forces off-setting the mesomeric enrichment in electrons of the o:p-positions; the electrostatic effect at the m-position is not compensated by the distribution of π-electrons, and the overall balance favours the p-position. With this model the first-order polarisability, taken into account by considering the self-atom polarisabilities $\pi_{r,\,r}$, parallels the polarisation in the ground state (cf. the case of quinoline; § 10.4.2), and according to it invoking the electromeric effect will not change the pattern predicted from the mesomeric effect, i.e. polarisability might have quantitative, but would not have qualitative, significance.[46]

Other substituents which belong with this group have already been discussed. These include phenol, anisole and compounds related to it (§ 5.3.4; the only kinetic data for anisole are for nitration at the encounter rate in sulphuric acid,[61] and with acetyl nitrate in acetic anhydride; see § 2.5 and § 5.3.3, respectively), and acetanilide (§ 5.3.4). The cations $PhSMe_2{}^+$, $PhSeMe_2{}^+$, and Ph_3O^+ have also been discussed (§ 9.1.2). Amino groups are prevented from showing their character $(-I\ +M)$ in nitration because conditions enforce reaction through the protonated forms (§ 9.1.2).

9.1.5 *Substituents containing boron or silicon*

Substituents containing boron are of interest because of the possibility which the boron atom offers of conjugation of a vacant orbital with the π-electrons of the benzene ring $(-M)$. The case of phenylboronic acid has been discussed (§ 5.3.4).

A silicon atom might be expected to release electrons inductively, but because of empty d-orbitals shows the overall character $(+I\ -M)$. Nitration of trimethylsilylbenzene[62] with nitric acid in acetic anhydride at -10 to $0\ ^\circ C$ gives 25·5, 39·8, 30·2 and 6·8 %, respectively, of o-, m-, and p-nitro-trimethylsilylbenzene and nitrobenzene, with a rate of reaction relative to that of benzene of about 1·5. The figures give no indication of an important conjugative effect.

Some results[63] for a series of compounds $Ph(CH_2)_n.SiMe_3$ $(n = 1\text{–}4)$, indicating enhancement of *o*-activity in the case ($n = 1$), have been explained by silicon–oxygen interaction in the transition state for *o*-substitution, as shown below.[57]

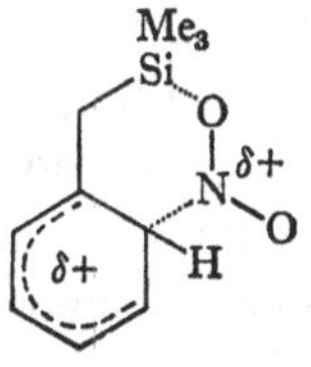

9.1.6 *Styryl and phenethynyl compounds*

The sparse data for this group of compounds are summarized in table 9.6. No partial rate factors are known, and a relative rate seems to have been determined only for cinnamic acid (0·11).[66] However, there is no reason to doubt that deactivation is the general condition. The substituents clearly resemble halogen substituents, achieving *o*:*p*-orientation by a polarizability of the double bond opposing the ground state polarization. This even obtains in what would appear to be the very unfavourable case of Ph.C(NO₂):CH.C₆H₄.NO₂(*p*) and also with Ph.CH:CH.NMe₃⁺, in which latter case the transition state contains adjacent positive charges.

TABLE 9.6 *The nitration of styryl and phenethynyl compounds*

Compound	Nitrating system	Isomer proportions /%			Ref.
		ortho	meta	para	
Ph.CH:CH.CO₂H	HNO₃	No *m*-substitution			64
Ph.CH:CH.NO₂	HNO₃, −15 °C	31	2	67	65
Ph.CH:CH.SO₂Cl	AcONO₂-Ac₂O, 25 °C	.	<2	.	66
Ph.CH:CH.NMe₃⁺	HNO₃	.	~2	.	67
Ph.C(NO₂):CH.C₆H₄. NO₂(*p*)	HNO₃, −20 to −15 °C	32	21	48	65
Ph.C:C.CO₂H	HNO₃, ⯦ −30 °C	27	8	65	49
Ph.C:C.CO₂Et	HNO₃, ⯦ −30 °C	36	6	58	49

9.2 DI- AND POLY-SUBSTITUTED DERIVATIVES OF BENZENE

There have been many studies of the orientation of nitration in di- and poly-substituted derivatives of benzene,[40, 56] but in very few cases have partial rate factors been determined. We shall be concerned mainly with these cases. So far as the data have permitted, interest has been centred round the 'additivity principle'. If this applied, two or more substituents on a benzene ring would each modify the Gibbs functions of activation for a particular position by the same amount as in the corresponding mono-substituted compound, resulting in an additive influence. The partial rate factor for a particular position in a di- or poly-substituted compound would therefore be given by the product of the appropriate partial rate factors for positions in mono-substituted compounds. It is not surprising that the principle is only moderately successful; it would, for example, be expected to fail where substituents interact strongly as when they are conjugated across the ring.

9.2.1 Di-substituted derivatives of benzene

The importance of a primary steric effect in the nitration of alkyl-benzenes has been mentioned (§9.1.1). The idea was first introduced by Le Fèvre[68a] to account for the fact that p-alkyltoluenes (alkyl = Et,[68b] i-Pr,[68a] t-Bu[68c]) are nitrated mainly adjacent to the methyl group. Without the rate data reported for the alkylbenzenes the effect might equally well have been accounted for by hyperconjugation.

The isomer proportions for the nitration of the chlorotoluenes, to be expected from the additivity principle, have been calculated[56] from the partial rate factors for the nitration of toluene and chlorobenzene and compared with experimental results for nitration with nitric acid at o °C. The calculated values are indicated in brackets beside the experimental values on the following structural formulae. In general, it can be

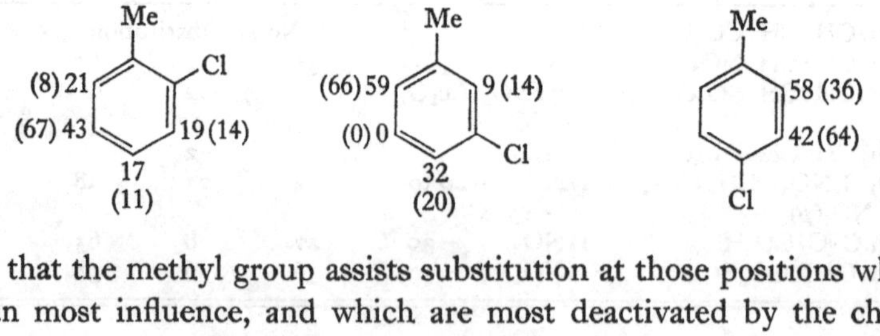

seen that the methyl group assists substitution at those positions which it can most influence, and which are most deactivated by the chloro substituent, more than would be predicted from its performance in toluene.

The same sort of situation is encountered in the nitration of the nitrotoluenes. The following diagrams record the observed partial rate factors ($10^8 f$) and, in parentheses, the corresponding calculated values derived from the partial rate factors for the nitration of toluene and of nitrobenzene,* and also the corresponding figures for relative rates.[41a] There is activation by the methyl group to a higher degree than is predicted. A deactivated nucleus produces a greater response from an activating substituent than does the phenyl group.

Me
(694) 11,600 — NO₂
0 — 124 (3·23)
23,200
(956)

Me
(21) 2740 — 3420 (51·3)
(41·2) 240 — NO₂
7540
(70·8)

Me
— 7920 (694)
— 15·8 (3·23)
NO₂

Relative rates (nitration in sulphuric acid at 25 °C)...

5170 × 10⁻⁸
(276 × 10⁻⁸)

2320 × 10⁻⁸
(30·7 × 10⁻⁸)

2640 × 10⁻⁸
(233 × 10⁻⁸)

Kinetic data are available for the nitration of a series of p-alkylphenyl trimethylammonium ions over a range of acidities in sulphuric acid.[70,71] The following table shows how p-methyl and p-*tert*-butyl augment the reactivity of the position *ortho* to them.[70] Comparison with table 9.1 shows how very much more powerfully both the methyl and the *tert*-butyl group assist substitution into these strongly deactivated cations than they do at the o-positions in toluene and *tert*-butylbenzene. Analysis of these results, and comparison with those for chlorination and bromination, shows that even in these highly deactivated cations, as in the nitration of alkylbenzenes (§9.1.1), the alkyl groups still release electrons in the inductive order.[70] In view of the comparisons just

Relative reactivities in Me₃N⁺—⟨ ⟩—R

	Me:H	*tert*-Bu:H
KNO₃/98 % H₂SO₄; −4 °C	3500	288
KNO₃/82 % H₂SO₄; 15·1 °C	2600	207

* Here, and with the chlorotoluenes, the precise values for the calculated figures depend on the values adopted for the partial rate factors in the mono-substituted compounds. These and the relative rates do depend slightly on conditions. As has been pointed out several times previously, comparisons with benzene for nitration in sulphuric acid have to be made with care.

made, it is interesting to note that a *p*-methyl group raises the rate of nitration of benzyl trimethyl ammonium about 360 times.[18]

Comparison of the rates of nitration in sulphuric acid of

$$p\text{-}X.\text{C}_6\text{H}_4.\text{NO}_2 \quad \text{and} \quad p\text{-}X.\text{C}_6\text{H}_4.\text{NMe}_3^+,$$

where X was a halogen or alkyl substituent, showed both of these kinds of substituent to reduce the ratio $k_{\text{NO}_2} : k_{\text{NMe}_3}^+$ from its value (1·9) for the case $X = \text{H}$. From being greater than unity for the halogen derivatives it becomes considerably less than unity for the alkyl derivatives. The substituents, both alkyl and halogen, assist substitution in the cation more than in the nitro-compound. This is attributed to polarizability effects, or equivalently, effects in the transition states. Because of the importance of the conjugation in the nitro-compounds a halogen substituent can stabilise the transition state less effectively than in the quaternary cation. Similarly hyperconjugation and induction produce a similar consequence for the alkyl compounds. Put another way, the greater polarisation of the *tert*-butyl group by nitro than by trimethylammonio reduces its ability to assist substitution by polarisability. The alkyl groups are more effective than halogen substituents in reducing the ratio $k_{\text{NO}_2} : k_{\text{NMe}_3}^+$ because with them this polarisability effect co-operates with the polarisation effect whereas

with the halogen substituents they are opposed. As regards the polarisation, the alkyl groups should reduce the ratio because more of the electron density which they supply remains on the ring in the quaternary cations than in the nitro-compounds, whilst the inductive effect of the halogen substituents should increase the ratio.[72]

Comparison of data for the nitration of alkyl- and halogenobenzenes with those for the related *p*-nitro-compounds supports the view that the rate of nitration of highly electron-deficient systems is determined by polarizability factors which enhance the reactivity of the substituted by comparison with that of the unsubstituted system.[72]

The suggestion outlined above about the way in which through-conjugation influences the nitration of *p*-chloronitrobenzene is relevant to the observed reactivities (*ortho* > *meta* > *para*) of the isomeric chloronitrobenzenes.[73] Application of the additivity principle to the

	X					
	Me	H	F	Cl	Br	I
Relative rate of overall nitration of Ph.X (AcONO$_2$, 18 °C)[72]	25	1·0	0·15	0·033	0·030	0·18
Relative rate of overall nitration of Ph.X (H$_2$SO$_4$, 25 °C)[46]	17	1·0	0·117	0·064	0·060	0·125
Relative rate of overall nitration of p-NO$_2$. C$_6$H$_4$.X (H$_2$SO$_4$, 25 °C)[72]	330	1·0	0·015	0·090	0·185	—

results for chloro- and nitro-benzene predicts the sequence *ortho* >
para > *meta*.[56]

Using the partial rate factors for nitration of chlorobenzene, Ridd and
de la Mare[56] calculated the relative rates of nitration of the dichlorobenzenes, with respect to p-dichlorobenzene, with the results shown
below. Also given are results based on more recent nitrations in mixed

	Relative rates of nitration				
Compound	HNO$_3$-AcOH, 20 °C	N$_2$O$_5$-CCl$_4$, 15 °C	Calc.[56]	HNO$_3$-H$_2$SO$_4$, 25 °C	Calc.
p-Dichlorobenzene	1·0	1·0	1·0	1·0	1·0
o-Dichlorobenzene	1·28	1·57	1·87	2·03	2·33
m-Dichlorobenzene	2·49	2·24	81·2	5·6	77·7

acid.[46] It was pointed out[56] that the additivity principle gave the correct
order of relative reactivities, but predicted that m-dichlorobenzene would
be relatively more reactive than it proved to be. This situation could
arise if o- and p-dichlorobenzene were more reactive than predicted;
that is, if the additivity principle underestimated the rate of substitution
at a position strongly deactivated by one substituent (a m-chloro
substituent) but less deactivated or activated by another (an o- or
p-chloro substituent). It could also arise if m-dichlorobenzene were much
less reactive than predicted. The details of the situation for nitration in
sulphuric acid emerge from the following diagrams, showing the observed ($10^4 f$) and calculated (in parentheses) partial rate factors.[46] In
m-dichlorobenzene both positions at which substitution occurs are less
reactive than predicted, the effect being most marked at the position

between the two chlorine atoms (where, presumably, a steric effect may operate). Conversely, positions deactivated by a *m*-chloro substituent are markedly assisted by an *o*-chloro substituent. The two opposing influences almost balance each other when one substituent is *meta* and the other *ortho* to the point of substitution.

Cl / Cl 4·1 (1·2)   Cl / Cl 31·9 (4·4)   Cl 7·9 (45) Cl 0·18   Cl 9·4 (16·5) Cl   Cl 8·8 (1·2) Cl

f_{obs}/f_{calc} 3·4   7·25   0·18   0·57   7·3

A M.O. treatment of the substituent effect of alkyl groups[25] has already been mentioned (§9.1.1). In this treatment, partial rate factors for the nitration of toluene and the xylenes with nitric acid in acetic acid at 25 °C were correlated with calculated cation localisation energies. A plot of $\log_{10} f$ against the difference between the cation localisation energy for a particular position and that for a position in benzene was linear and passed through the origin. The points falling furthest from the line were those for the 4 position in *o*-xylene and the 2 position in *m*-xylene. Both of these sites were less reactive than predicted, and for the 2-position in *m*-xylene this was put down to steric compression. Even so, the success of the correlation is surprising, as the following considerations show. The diagrams give the observed partial rate

Me / Me 61·4 (118·1) 23·3 (166·6)   Me 268 (2420·6) 1063 (3414·5) Me   Me 54·9 (118·1) Me

Relative rate (nitration in acetic acid at 25 °C)...

28·2 ± 3·8 (94·9)   399 ± 48 (1541·6)   36·6 (78·7)

factors[25] for the several positions in the xylenes, and, in parentheses, those calculated from appropriate data for toluene[5] (table 9.1), as well as the observed[25] and calculated relative rates. It is clear that all of the xylenes react more slowly than expected, and certain that reaction upon encounter is affecting the results, especially in the case of *m*-xylene.

9.2.2   *Poly-substituted derivatives of benzene*

Some observations about the nitration of some polymethylbenzenes have already been made (§§6.3, 6.4). Replacement of an alkyl group has frequently been observed,[74] but quantitative studies are lacking.

Table 9.7 contains recent data on the nitration of polychlorobenzenes in sulphuric acid. The data continue the development seen with the dichlorobenzenes. The introduction of more substituents into these deactivated systems has a smaller effect than predicted. Whereas the *p*-position in chlorobenzene is four times less reactive than a position in benzene, the remaining position in pentachlorobenzene is about four times more reactive than a position in 1,2,4,5-tetrachlorobenzene. The chloro substituent thus activates nitration, a circumstance recalling the fact that *o*-chloronitrobenzene is more reactive than nitrobenzene.[42b] As can be seen from table 9.7, the additivity principle does not work very well with these compounds, underestimating the rate of reaction of pentachlorobenzene by a factor of nearly 250, though the failure is not so marked in the other cases, especially viewed in the circumstance of the wide range of reactivities covered.

TABLE 9.7   *The nitration of polychlorobenzenes in sulphuric acid*[46]* *at 25 °C*

Compound	Relative rate†	Partial rate factors		
		f_{obs}	f_{calc}	f_{obs}/f_{calc}
1,3,5-Trichlorobenzene	$5 \cdot 5 \times 10^{-4}$	$0 \cdot 0011$	$0 \cdot 0011$	$1 \cdot 0$
1,2,3,4-Tetrachlorobenzene	$1 \cdot 6 \times 10^{-6}$	$4 \cdot 8 \times 10^{-6}$	$5 \cdot 3 \times 10^{-8}$	$9 \cdot 0$
1,2,3,5-Tetrachlorobenzene	$4 \cdot 0 \times 10^{-6}$	$8 \cdot 0 \times 10^{-6}$	$2 \cdot 0 \times 10^{-6}$	$4 \cdot 0$
1,2,4,5-Tetrachlorobenzene	$6 \cdot 9 \times 10^{-8}$	$2 \cdot 0 \times 10^{-7}$	$1 \cdot 45 \times 10^{-8}$	$13 \cdot 8$
Pentachlorobenzene	$1 \cdot 4 \times 10^{-7}$	$8 \cdot 8 \times 10^{-7}$	$3 \cdot 6 \times 10^{-9}$	244

* For all of the compounds other than 1,2,3,4,-tetrachlorobenzene rates at more than one acidity are reported.
† Relative to that of benzene.

The nitration of some substituted nitrobenzenes has been studied in connection with the high *o*:*p*-ratios produced by $[-I-M]$ substituents. Thus nitration in sulphuric acid of 2,5-dialkyl-nitrobenzenes produces the isomer distributions shown below.[75] As has been seen (§9.1.3), one explanation for the occurrence of high *o*:*p*-ratios with $[-I-M]$ substituents is that the latter specifically deactivate *para* positions. In the present cases the twisting of the nitro-group by the adjacent alkyl group should reduce any such selective deactivation. The fact that despite this the *o:p*-ratio in these cases remains greater than unity, and is not sensitive to a change of alkyl group, is thought to be best explained by the unequal distribution of positive charge at the nuclear positions in the transition state (§9.1.3).

				o:p-ratio
$R = H$ (%)	6·12	2·06	91·8	3·0
Me	48	12	40	4·0
Et	52	19	29	2·7
tert-Bu	.	.	Predominant	.

A similar study of the nitration of 2,5-dichloro- and 2,5-dibromo-nitrobenzene under a variety of conditions has been made.[76] At the very high acidities in oleum the *o:p*-ratio for nitration was less than unity. It increased with decreasing acidity of the medium and became greater than unity at roughly the acidity represented by 89–90% sulphuric acid. The results were interpreted in terms of the interaction between the nitronium ion and the nitro group, but the results are complicated and the interpretation not compelling.

9.3 HETEROCYCLIC COMPOUNDS

There is available a large amount of qualitative information about the nitration of heterocyclic compounds, but quantitative information is still not very extensive, being limited to nitrogen systems.

9.3.1 *Pyridine derivatives*

For this series of compounds qualitative information is quite extensive.[77] Application of the criteria discussed in §8.2, in particular comparison with the corresponding methyl quaternary salt, establishment of the rate profile for nitration in sulphuric acid, and consideration of the encounter rate and activation parameters, shows that 2,4,6-collidine is nitrated as its cation.[78,79a] The same is true for the 3-nitration of 2,4-[79a]

and 2,6-dimethoxypyridine,[79] and the 2-nitration of 3,5-dimethoxy-pyridine.[80] In contrast, such criteria show that the conversion of 2,6-dimethoxy-3-nitro-[79] and 3,5-dimethoxy-2-nitro-[80] into 2,6-dimethoxy-3,5-dinitro- and 3,5-dimethoxy-2,6-dinitro-pyridine, respectively, proceeds through the free bases. The same is true for the formation of 2,6-dichloro-3-nitropyridine from 2,6-dichloropyridine.[79a] Broadly, pyridine derivatives with $pK_a > +1$ will be nitrated as their cations and nitration will occur at α- or β-positions depending upon the orientation of activating substituents which may be present; whilst pyridine derivatives with $pK_a < -2\cdot5$ will be nitrated as the free bases.[80]

The precise comparison of the effects of substituents upon the reactivity of a pyridine or pyridinium nucleus with their effects in a benzene nucleus, and similarly, of the effect of the hetero-atom upon the reactivity of the benzene nucleus, is complicated by several factors. There is first the difficulty, mentioned already at several points in this book, that data for the deactivated pyridine derivatives will be obtained by measurements made at high acidities, and also, in some cases, including some of those discussed above, at relatively high temperatures. Then there may be the problem of choosing an acidity scale which measures the degree of protonation of the base being considered. Finally, it may not be possible to compare the rate of nitration of a pyridine derivative with that of the corresponding benzene derivative because nitration of the latter occurs at the encounter rate; this is the case when it is attempted to compare the dimethoxypyridines with *m*-dimethoxybenzene.[79a–80] Some of these difficulties are encountered in estimating the partial rate factor of about 10^{-12} for the 2,4,6-trimethylpyridinium ion.[79a] The very strong deactivating effect of the $:\overset{+}{\text{N}}\text{H}$ implied by this figure is consonant with the fact that the value of the substituent constant (σ) for this group is the largest known.[77]

The similarity of their rate profiles, and the similarity of their rate constants for nitration at a particular temperature and acidity show that 4-pyridone, 1-methyl-4-pyridone, and 4-methoxypyridine are all nitrated as their cations down to about 85 % sulphuric acid.* The same is true of 2-methoxy-3-methylpyridine. In contrast, 3- and 5-methyl-2-pyridone, 1,5-dimethyl-2-pyridone and 3-nitro-4-pyridone all react

---

* 4-Pyridone is one of the compounds for which the encounter rate criterion for choosing between nitration *via* the free base or by the conjugate acid is ambiguous (§8.2.3).

as the free bases, and for the first two compounds the free base is probably in the pyridone, rather than the hydroxypyridine form.[81]

At lower acidities (65–86 % sulphuric acid) the 2-pyridones mentioned continue to be nitrated as the free bases, and despite some uncertainty arising from the fact that the nitrations were carried out at relatively high temperatures (85–157·5 °C) (§8.2.1), it is probable that in this region of acidity 4-pyridone also reacts as the free base.[81] Data for the acidity dependence of their rates of nitration have been given for these pyridones in table 8.2. Evidently for pyridones nitration occurs through the free base if $pK_a < 1·5$; 4-pyridone is more basic (pK_a 3·27) and in strongly acidic media reacts as its cation as already described. The 2-pyridones undergoing nitration as the free bases are deactivated by a factor of about 10 at the β-positions, whilst for the cations of 4-pyridone and 2- and 4-methoxypyridine deactivation by a factor not less than 10^{13}, as compared with anisole (which reacts at the encounter rate), is indicated.

The 2-nitration of 3-hydroxy- and 3-methoxy-pyridine in 85–96 % sulphuric acid involves the conjugate acids, whilst the 3-nitration of 6-hydroxy and 6-methoxy-2-pyridone in 70–77 % sulphuric acid involves the free bases, which react at, or near to the encounter rate.[79b]

The interest attaching to the nitration of pyridine 1-oxide and its derivatives has already been mentioned (§8.2.5). Some data for these reactions are given in tables 8.1, 8.2 and 8.4. The 4-nitration of pyridine 1-oxide is shown to occur through the free base by comparison with the case of 1-methoxypyridinium cation (§8.2.2), by the nature of the rate profile (§8.2.1), and by consideration of the encounter rate (§8.2.3).[82a, 83] Some of these criteria have been used to show that the same is true for 2,6-[82a] and 3,5-lutidine 1-oxide,[83] and for 2,6- and 3,5-dichloropyridine 1-oxides.[83] These nitrations were necessarily carried out at temperatures appreciably higher than 25 °C.

The more basic and reactive compounds, 4-methoxy-2,6-dimethyl-, 2,6-dimethoxy-, 3,5-dimethoxy and 2,4,6-trimethoxy-pyridine 1-oxide are nitrated at convenient speeds at temperatures near to 25 °C. The reactions involve the conjugate acids, and substitution occurs at $C_{(3)}$, except in the case of 3,5-dimethoxypyridine 1-oxide which reacts at $C_{(2)}$.* The further nitration of 3,5-dimethoxy-2-nitropyridine 1-oxide to give the 2,6-dinitro compound is more difficult but it seems likely that it also involves the conjugate acid.[83]

Any attempt to calculate partial rate factors for the 1-oxides being

* The encounter rate criterion (§8.2.3) is ambiguous for these compounds.

nitrated as the free bases faces all of the difficulties recalled in the discussion of pyridines above, and the magnitude of these difficulties is shown in the vastly different values which have been calculated for $C_{(4)}$ in pyridine 1-oxide itself $(2 \cdot 1 \times 10^{-3}$ and $4 \times 10^{-6})$.[82a, 83] In one case Arrhenius extrapolation to 25 °C was followed by correction to figures relating to the free base by use of the factor h_0/K_a and then by comparison with data for the quinolinium ion, whilst in the other the factor h_a/K_a was used and comparison made with data for benzene extrapolated to 87·9 % sulphuric acid. In the first case h_0 was used because that acidity function brings the rate profile for the nitration of 2,6-lutidine 1-oxide better into relationship with normal rate profiles than does h_A,[84] in the other h_A was used because it better represents the protonation of oxides than does h_0.[83]

With the oxides which are nitrated as the cations the difficulties are much less serious for the use of an acidity function is not involved. Comparison of 2,6-dimethoxy- and 3,5-dimethoxy-pyridine 1-oxide with *m*-dimethoxybenzene (which is nitrated at the encounter rate) shows that in these cases the $\overset{+}{:\mathrm{N}}\text{-OH}$ group deactivates the ring by factors greater than 10^7 and 10^8, respectively.[83]

Numerous M.O.-theoretical studies have been made of reactivity indices (§7.2.2, 7.2.3) relevant to substitution into pyridine and pyridinium. Electron densities have mostly been calculated using the Hückel approximation, though more advanced methods have been used;[77, 85] they are somewhat ambiguous in their implications for electrophilic substitution.[77, 85a] Localisation energies predict reasonably well the broad facts of orientation, but are not so successful in representing the state of de-activation of the heterocyclic nuclei.[77] The Hückel approximation correctly predicts that electrophilic substitution into pyridine 1-oxide should occur at $C_{(4)}$ in the free base and at $C_{(3)}$ in the conjugate acid, but again performs badly in describing the reactivities of the nuclear positions, predicting partial rate factors greater than unity.[82] A more recent treatment of electrophilic substitution in pyridine 1-oxide attempts to make allowance for the differing characters of electrophiles.[86]

9.3.2 *Azoles*

The kinetics of nitration in sulphuric acid of both pyrazole and imidazole have been studied.[87] Data have already been quoted (tables 8.1, 8.3) to support the view that the nitration of both of these compounds at $C_{(4)}$

involves the conjugate acids. The rate profile for nitration of pyrazole below 90 % sulphuric acid leaves a slight doubt on this point, and the case of imidazolium is complicated by the dependence upon acidity of the yield of 4-nitroimidazole. Partial rate factors for the 4-nitration of pyrazolium and imidazolium in 98 % sulphuric acid at 25 °C were calculated to be $2 \cdot 1 \times 10^{-10}$ and $3 \cdot 0 \times 10^{-9}$, respectively.

M.O. theory has had limited success in dealing with electrophilic substitution in the azoles. The performances of π-electron densities as indices of reactivity depends very markedly on the assumptions made in calculating them.[85, 88] Localisation energies have been calculated for pyrazole and pyrazolium, and also an attempt has been made to take into account the electrostatic energy involved in bringing the electrophile up to the point of attack; the model predicts correctly the orientation of nitration in pyrazolium.[88]

9.4 $\rho^+ \sigma^+$ CORRELATIONS IN NITRATION

The development of linear free energy correlations of the rate of aromatic substitutions has been discussed (§7.3). We record here the results of such correlations for nitration.

For the nine substituents *m*- and *p*-methyl, *p*-fluoro, *m*- and *p*-chloro, *m*- and *p*-bromo, and *m*- and *p*-iodo, using the results for nitration carried out at 25 °C in nitromethane or acetic anhydride[1, 48, 58–9] (see tables 9.1, 9.5), a plot of $\log_{10} k/k_0$ against σ^+ produced a substituent constant $\rho = -6 \cdot 53$ with a standard deviation from the regression line $s = 0 \cdot 335$, and a correlation coefficient $c = 0 \cdot 975$.[89a] Inclusion of results for *m*- and *p*-ethoxycarbonyl[48] (see table 9.4) and for *p*-phenyl (see §10.1), some of which referred to 0° C, gave $\rho = -6 \cdot 22$ ($s = 0 \cdot 287$; $c = 0 \cdot 980$).[89b] Fig. 9.1 is a plot[90] which also includes data for 2-fluorenyl and β-naphthyl (see §§10.1, 10.2).

Considering that the results used for these plots relate to nitrations carried out under different conditions, the success of the correlations is remarkable.

The Yukawa–Tsuno equation [$\log k/k_0 = \rho\{\sigma + r(\sigma^+ - \sigma)\}$] (§7.3.1) applied to nitration at 25 °C in nitromethane or acetic acid[91] gives $\rho = -6 \cdot 38$, $r = 0 \cdot 90$.

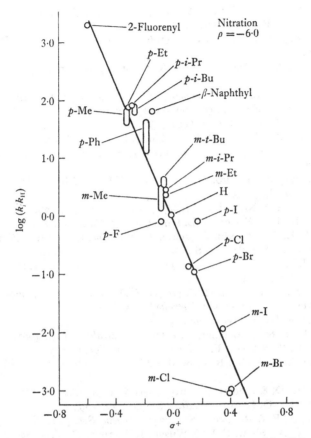

Fig. 9.1. The relationship between the σ^+ constants and log (k/k_H) for nitration. (From Stock & Brown.[90])

REFERENCES

1. Ingold, C. K., Lapworth, A., Rothstein, E. & Ward, D. (1931). *J. chem. Soc.* p. 1959.
2. Knowles, J. R., Norman, R. O. C. & Radda, G. K. (1960). *J. chem. Soc.* p. 4885.
3. Knowles, J. R. & Norman, R. O. C. (1961). *J. chem. Soc.* p. 2938.
4. Stock, L. M. (1961). *J. org. Chem.* **26**, 4120.
5. Olah, G. A., Kuhn, S. J., Flood, S. H. & Evans, J. C. (1962). *J. Am. chem. Soc.* **84**, 3687.
6. Cohn, H., Hughes, E. D., Jones, M. H. & Peeling, M. G. (1952). *Nature, Lond.* **169**, 291.
7. Brown, H. C. & Wirkkala, R. A. (1966). *J. Am. chem. Soc.* **88**, 1447.
8. Ingold, C. K. & Ingold, E. H. (1928). *J. chem. Soc.* 2249.

References

9. Ingold, C. K., Ingold, E. H. & Shaw, F. R. (1927). *J. chem. Soc.* p. 813.
10. Holleman, A. F. (1925). *Chem. Rev.* 1, 187.
11. Albers, R. J. & Kooyman, E. C. (1964). *Recl Trav. chim. Pays-Bas Belg.* 83, 930.
12. Coombes, R. G., Moodie, R. B. & Schofield, K. (1969). *J. chem. Soc.* B, p. 52.
13. Ridd, J. H. (1964). *Nitro Compounds*, p. 43. Proceedings of the International Symposium, Warsaw, 1963. London: Pergamon Press.
14. Brickman, M. & Ridd, J. H. (1965). *J. chem. Soc.* p. 6845.
15. Brickman, M., Utley, J. H. P. & Ridd, J. H. (1965). *J. chem. Soc.* p. 6851.
16. Hartshorn, S. R. & Ridd, J. H. (a) (1968). *J. chem. Soc.* B, p. 1063. (b) p. 1068.
17. Gastaminza, A., Modro, T. A., Ridd, J. H. & Utley, J. H. P. (1968). *J. chem. Soc.* B, p. 534.
18. Modro, T. A. & Ridd, J. H. (1968). *J. chem. Soc.* B, p. 528.
19. Gilow, H. M. & Walker, G. L. (1967). *J. org. Chem.* 32, 2580.
20. Riley, F. L. & Rothstein, E. (1964). *J. chem. Soc.* (a) p. 3860; (b) p. 3872.
21. Ingold, C. K. & Wilson, I. S. (1927). *J. chem. Soc.* p. 810.
22. Baker, J. W. & Moffitt, W. G. (a) (1930). *J. chem. Soc.* p. 1722; (b) (1931). *J. chem. Soc.* p. 314.
23. Nelson, K. L. & Brown, H. C. (1951). *J. Am. chem. Soc.* 73, 5605.
24. de la Mare, P. B. D. (1959). *Tetrahedron* 5, 107. Berliner, E. (1959). *Tetrahedron* 5, 202.
25. Clark, D. T. & Fairweather, D. J. (1969). *Tetrahedron* 25, 4083.
26. Ingold, C. K. & Shaw, F. R. (1949). *J. chem. Soc.* p. 575.
27. Roberts, J. D., Webb, R. L. & McElhill, E. A. (1950). *J. Am. chem. Soc.* 72, 408.
28. Vorländer, D. & Siebert, E. (1919). *Ber. dt. chem. Ges.* 52, 283. Vorländer, D. (1925). *Ber. dt. chem. Ges.* 52, 1893.
29. Challenger, F. & Rothstein, E. (1934). *J. chem. Soc.* p. 1258.
30. Ingold, C. K., Shaw, F. R. & Wilson, I. S. (1928). *J. chem. Soc.* p. 1280.
31. Sandin, R. B., McClure, F. T. & Irwin, F. (1939). *J. Am. chem. Soc.* 61, 3061.
32. Goss, F. R., Hanhart, W. & Ingold, C. K. (1927). *J. chem. Soc.* p. 250.
33. Ingold, C. K. (1953). *Structure and Mechanism in Organic Chemistry*, ch. 6. London: Bell.
34. Goss, F. R., Ingold, C. K. & Wilson, I. S. (1926). *J. chem. Soc.* p. 2440.
35 Nesmeyanov, A N., Tolstaya, T. P., Isaeva, L. S. & Grib, A. V. (1960). *Dokl. Akad. Nauk. SSSR* 133, 602.
36. Roberts, J. D., Clement, R. A. & Drysdale, J. J. (1951). *J. Am. chem. Soc.* 73, 2181.
37. Ridd, J. H. (1967). *Spec. Publs Chem. Soc.* 21, 149.
38. Chandra, A. K. & Coulson, C. A. (1965). *J. chem. Soc.* p. 2210.
39. Bishop, D. M. & Craig, D. P. (1963). *Molec. Phys.* 6, 139.
40. Holleman, A. F. (1910). *Die direkte Einführung von Substituenten in den Benzolkern.* Leipzig: Veit.
41. (a) Tillett, J. G. (1962). *J. chem. Soc.* p. 5142. (b) Mésure, A. D. & Tillett, J. G. (1966). *J. chem. Soc.* B, p. 669.

42. (*a*) Westheimer, F. H. & Kharasch, M. S. (1946). *J. Am. chem. Soc.* **68**, 1871.
   (*b*) Vinnik, M. I., Grabovskaya, Zh. E. & Arzamaskova, L. N. (1967). *Russ. J. phys. Chem.* **41**, 580.
43. Bonner, T. G., James, M. E., Lowen, A. M. & Williams, G. (1949). *Nature, Lond.* **163**, 955.
44. Aliprandi, B., Cacace, F. & Ciranni, G. (1964). *Analyt. Chem.* **36**, 2445.
45. Moodie, R. B., Penton, J. R. & Schofield, K. (1969). *J. chem. Soc.* B, p. 578.
46. Coombes, R. G., Crout, D. H. G., Hoggett, J. G., Moodie, R. B. & Schofield, K. (1960). *J. chem. Soc.* B, p. 347.
47. Cooper, K. E. & Ingold, C. K. (1927). *J. chem. Soc.* p. 836.
48. Ingold, C. K. & Smith, M. S. (1938). *J. chem. Soc.* p. 905.
49. Baker, J. W., Cooper, K. E. & Ingold, C. K. (1928). *J. chem. Soc.* p. 426.
50. Wibaut, J. P. & van Strik, R. (1958). *Recl. Trav. chim. Pays-Bas Belg.* **77**, 316.
51. Twist, R. F. & Smiles, S. (1925). *J. chem. Soc.* **127**, 1248.
52. Obermiller, J. (1914). *J. prakt. Chem.* **69**, 70.
53. Lapworth, A. & Robinson, R. (1928). *Mem. Proc. Manch. lit. phil. Soc.* **72**, 243.
54. Roberts, J. D. & Streitwieser, A. (1952). *J. Am. chem. Soc.* **74**, 4723.
   Brown, R. D. (1953). *J. Am. chem. Soc.* **75**, 4077.
55. Ingold, C. K. (1926). *A. Rep. chem. Soc.* **23**, 140.
56. Ridd, J. H. & de la Mare, P. B. D. (1959). *Aromatic Substitution: Nitration and Halogenation*, ch. 6. London: Butterworths.
57. Norman, R. O. C. & Taylor, R. (1965). *Electrophilic Substitution in Benzenoid Compounds.* London: Elsevier.
58. Roberts, J. D., Sanford, J. K., Sixma, F. L. J., Cerfontain, H. & Zagt, R. (1954). *J. Am. chem. Soc.* **76**, 4525.
59. Bird, M. L. & Ingold, C. K. (1938). *J. chem. Soc.* p. 918.
60. de la Mare, P. B. D. & Robertson, P. W. (1948). *J. chem. Soc.* p. 100.
61. Deno, N. C. & Stein, R. (1956). *J. Am. chem. Soc.* **78**, 578.
62. Speier, J. L. (1953). *J. Am. chem. Soc.* **75**, 2930.
63. Chernyshev, E. A., Dolgaya, H. E. & Petrov, A. D. (1960). *Izv. Akad. Nauk SSSR, Otd. Khim. nauk.* p. 1424.
64. Underwood, H. W. & Kochman, E. L. (1926). *J. Am. chem. Soc.* **48**, 264.
65. Baker, J. W. & Wilson, I. S. (1927). *J. chem. Soc.* p. 842.
66. Bordwell, F. G. & Rohde, K. (1948). *J. Am. chem. Soc.* **70**, 1191.
67. Truce, W. E. & Simms, J. A. (1957). *J. org. Chem.* **22**, 762.
68. (*a*) Le Fèvre, R. J. W. (1933). *J. chem. Soc.* p. 980.
   (*b*) Brady, O. L. & Day, J. N. E. (1934). *J. chem. Soc.* p. 114.
   (*c*) Battegay, M. & Haeffely, P. (1924). *Bull. Sci. chim. France* **35**, 981.
69. Wibaut, J. P. (1913). *Recl Trav. chim. Pays-Bas Belg.* **32**, 244.
70. Utley, J. H. P. & Vaughan, T. A. (1968). *J. chem. Soc.* B, p. 196.
71. Williams, G. & Lowen, A. M. (1950). *J. chem. Soc.* p. 3312.
   Lowen, A. M., Murray, M. A., & Williams, G. (1950). *J. chem. Soc.* p. 3318.
72. Brand, J. C. D. & Paton, R. P. (1952). *J. chem. Soc.* p. 281.
73. Martinsen, H. (1907). *Z. phys. Chem.* **59**, 605.

References

74. Nightingale, D. V. (1947). *Chem. Rev.* **40**, 117. See also ch. 16 of ref. 56. above.
75. Johnson, C. D. & Northcott, M. J. (1967). *J. org. Chem.* **32**, 2029.
76. Hammond, G. S., Modic, F. J. & Hedges, R. M. (1953). *J. Am. chem. Soc.* **75**, 1388.
77. Schofield, K. (1967). *Hetero-Aromatic Nitrogen Compounds: Pyrroles and Pyridines.* London: Butterworth.
78. Katritzky, A. R. & Ridgewell, B. J. (1963). *J. chem. Soc.* p. 3882.
79. (a) Johnson, C. D., Katritzky, A. R., Ridgewell, B. J., & Viney, M. (1967). *J. chem. Soc.* B, p. 1204.
    (b) Katritzky, A. R., Tarhan, H. O. & Tarhan, S. (1970). *J. chem. Soc.* B, p. 114.
80. Johnson, C. D., Katritzky, A. R. & Viney, M. (1967). *J. chem. Soc.* B, p. 1211.
81. Brignell, P. J., Katritzky, A. R. & Tarhan, H. O. (1968). *J. chem. Soc.* B, p. 1477.
82. (a) Gleghorn, J., Moodie, R. B., Schofield, K. & Williamson, M. J. (1966). *J. chem. Soc.* B, p. 870.
    (b) Gleghorn, J. (1967). Ph.D. thesis, University of Exeter.
83. Johnson, C. D., Katritzky, A. R., Shakir, N. & Viney, M. (1967). *J. chem. Soc.* B, p. 1213.
84. Gleghorn, J. T., Moodie, R. B., Qureshi, E. A. & Schofield, K. (1968). *J. chem. Soc.* B, p. 316.
85. (a) Ridd, J. H. (1963). In *Physical Methods in Heterocyclic Chemistry* (vol. 1), ed. A. R. Katritzky. vol. 1. New York: Academic Press.
    (b) Adam, W., Grimison, A. & Rodriguez, G. (1967). *Tetrahedron* **23**, 2513.
86. Klopman, G. (1968). *J. Am. chem. Soc.* **90**, 223.
87. Austin, M. W., Blackborow, J. R., Ridd, J. H. & Smith, B. V. (1965). *J. chem. Soc.* p. 1051.
88. Finar, I. L. (1968). *J. chem. Soc.* B, p. 725.
89. Brown, H. C. & Okamoto, Y. (a) (1957). *J. Am. chem. Soc.* **79**, 1913; (b) (1958). *J. Am. chem. Soc.* **80**, 4979.
90. Stock, L. M. & Brown, H. C. (1963). *Adv. phys. org. Chem.* **1**, 35.
91. Yukawa, T. & Tsuno, Y. (1959). *Bull. chem. Soc. Japan* **32**, 971.

10 Nitration and aromatic reactivity: D. The nitration of bi- and poly-cyclic compounds

10.1 BIPHENYLS AND RELATED SYSTEMS

Despite the considerable amount of work which has been reported, our knowledge of the nitration of biphenyl is not in a satisfactory state. The $o:p$-ratio varies considerably with the conditions of nitration, and the cause of the variation is not fully understood. Nitrations with solutions prepared from nitric acid and acetic anhydride have generally given $o:p$-ratios greater than unity,[1a–4] the most consistent value[2,4] being 2·2, obtained at 0 °C. The corresponding partial rate factors are reported later.

In contrast, nitration with nitric acid in acetic acid,[5] with nitric acid in sulphuric acid,[6] and with liquid dinitrogen tetroxide[7] gave $o:p$-ratios of 0·5–0·6. Some of the more recent results are collected in table 10.1. The first five entries appeared to reveal a major difference between nitration in homogeneous and in heterogeneous systems, the former type giving higher $o:p$-ratios than the latter. In homogeneous systems biphenyl does not behave like anisole or acetanilide (§5.3.4), the changes of orientation with medium being small. The variation in $o:p$-ratio was ascribed to stereochemical differences existing between biphenyl in solution and solid biphenyl.[4,8] The regularity of this situation is disturbed by later results[9] (table 10.1) which re-affirm the older finding that nitration in acetic acid gives an $o:p$-ratio of 0·6.

The effect of heterogeneity upon nitration in sulphuric acid was examined by Taylor,[10] who did nitrations at temperatures between 0 and 70 °C both with and without nitrobenzene present as solvent for the biphenyl. He concluded that an $o:p$-ratio of $\sim$ 0·6 for nitration in sulphuric acid arose from an 'essentially homogeneous reaction', heterogeneous nitration occurring only at lower temperatures. Thus, it seems to be established that homogeneous conditions of nitration, except those using solutions prepared from nitric acid and acetic anhydride, give low $o:p$-ratios, and no special explanation for these ratios (see above) is required. The outstanding problem is to deduce the nature of the

TABLE 10.1   *The nitration of biphenyl*

Reagent	Temp./°C	Isomer proportions/(%)		Ref.
		2	*4*	
HNO₃–Ac₂O	0	69·2	30·8	4
HNO₃–AcOH	0	64·4	35·6	4
1:5 HNO₃–H₂SO₄ in PhNO₂	0	65·0	35·0	4
HNO₃–H₂SO₄ (heterogeneous)	0	42·3	57·7	4
	35	43·1	56·9	4
		o:p-Ratio		
HNO₃–Ac₂O	0–18	2·3		9
HNO₃–AcOH	85–90	0·6		9
HNO₃–Ac₂O	−40	2·74		10
	0	2·15		10
	25	1·96		10
N₂O₅–MeCN	−20	2·79		10
	0	2·67		10
	25	2·58		10

electrophile in solutions of nitric acid in acetic anhydride, and then to account for the high *o:p*-ratio which it gives. From the similarity of the results obtained using solutions prepared from nitric acid and acetic anhydride and solutions of dinitrogen pentoxide in acetonitrile (table 10.1), and in the light of the evidence for protonated acetyl nitrate being the electrophile in the former solutions (§5.3), Taylor argued that nitration of biphenyl with nitronium ions gives normal *o:p*-ratios, whilst reagents of the type NO₂X (e.g. protonated acetyl nitrate or dinitrogen pentoxide) give high *o:p*-ratios. He adapted the special mechanisms proposed to explain the high *o:p*-ratios obtained with ethers and anilides (§5.3.4) to the case of biphenyl by supposing the π-electrons of the ring not being nitrated to interact with the reagent NO₂X; the resulting π-complex would favour o-substitution and the observed *o:p*-ratios would depend upon the relative proportions of nitration by this mechanism and by the nitronium ion mechanism. The results of nitration with nitronium tetrafluoroborate have been included in this scheme.[11]

Taylor[10] also reinvestigated the isomer proportions resulting from the nitration of biphenyl with nitric acid in acetic acid. The result obtained depended markedly upon the composition of the nitrating solution and upon the way in which nitration was carried out. Thus, adding nitric acid to biphenyl dissolved in acetic acid at 85–90 °C gave an $o:p$-ratio of 0·7, but rapid stirring increased the time of reaction fivefold and the $o:p$-ratio to 1·2. The former result was attributed to nitration by nitronium ions, formed by self-protonation of nitric acid present in high local concentrations, and the latter to the operation of another electrophile, possibly molecular nitric acid or acetyl nitrate produced from nitric acidium ion and acetic acid. It has not hitherto been necessary to postulate the occurrence of any but the nitronium ion mechanism for reactions in acetic acid (§ 3.2), and these results need further investigation.

It should be noted that reported kinetic data on the nitration of biphenyl are limited to one rate constant for reaction in 68·3 % sulphuric acid at 25 °C (k_2 = 0·92 l mol^{-1} s^{-1}; relative rate, 15·8).[12] Until they have been extended the above discussion must be regarded as provisional.

Much work has been done to determine the effect upon the orientation of nitration of substituents in biphenyl. 4-Methylbiphenyl with nitric acid in acetic acid gives mainly the 4'- and 2'-nitro-derivatives with some of the 2-nitro-derivative,[13a] whilst 3-methyl- and 3,5-dimethyl-biphenyl give the 4-nitro-derivatives.[13b–d] Nitration of 2-acetylamino-biphenyl with nitric acid in acetic acid gives 2-acetylamino-5-nitro-biphenyl,[14] whilst nitration in a mixture of acetic acid and sulphuric acid gives 2-acetylamino-4'-nitrobiphenyl;[15a] both of these mononitro-compounds are further nitrated to give 2-acetylamino-4',5-dinitro-biphenyl.[14, 15a] The reagent used also influences the orientation of nitration of 3-acetylaminobiphenyl; a reagent prepared from acetic acid, acetic anhydride, and nitric acid gives 3-acetylamino-4-nitrobiphenyl, whilst nitric acid alone gives 3-acetylamino-4'-nitrobiphenyl.[16] Nitration of 4-acetylaminobiphenyl with nitric acid gives 4-acetylamino-3-nitro- and 4-acetylamino-3,4'-dinitrobiphenyl.[15b] 2-p-Toluenesulphonyl-aminobiphenyl with aqueous nitric acid is nitrated at the 5-position, and with nitric acid in acetic acid at the 3,5-positions, whilst with the latter reagent 4-p-toluenesulphonylaminobiphenyl gives the 3,5-dinitro-compound.[14]

Aromatic reactivity: D Bi- and poly-cyclic compounds

The varying behaviour of the acetylamino group with different nitrating media may well be due to its protonation in more strongly acidic solutions. Otherwise, the situation appears to be that a weakly activating group when most favourably placed to do so, i.e. when at $C_{(4)}$ will reinforce the $o:p$-directing effect of phenyl, leading to $4'$-nitration. Such a group in any other position, or a more powerfully activating group at any position, directs nitration *ortho* or *para* to itself. The great loss of intensity suffered by the electronic effects of substituents, when these effects are transmitted across the bond linking the two phenyl groups,[17a] makes these substituted biphenyls best regarded as disubstituted benzene derivatives.

The nitration of nitro- and dinitro-biphenyls has been examined by several workers.[2, 16, 18] As would be expected, nitration of the nitro-biphenyls occurs in the phenyl ring. Like a phenyl group, a nitrophenyl group is $o:p$-directing, but like certain substituents of the type ·CH: CHX (§9.1.6) it is, except in the case of *m*-nitrophenyl, deactivating. Partial rate factors for the nitration at o °C of biphenyl and the nitro-biphenyls with solutions prepared from nitric acid and acetic anhydride are given below. The high $o:p$-ratio found for nitration of biphenyl

	Ph	$C_6H_4.NO_2(o)$	$C_6H_4NO_2(m)$	$C_6H_4.NO_2(p)$
ortho	41 (36·4)	0·28	1·4	0·35
meta	< 0·6	0·03	0	< 0·01
para	38 (32·6)	1·2	3·4	1·3
$o:p$-Ratio ...	2·2	0·46	0·82	0·54
Relative rate	41	0·3	1·03	0·33

Partial rate factors are from ref. 2 and (in parentheses) ref. 4. $o:p$-Ratios are from ref. 2 and relative rates are calculated from the partial rate factors. For biphenyl Dewar et al.[1d–f] give partial rate factors $f_2 = 30$; $f_4 = 18$. Recalculation of their results gives $f_2 = 18·7$; $f_4 = 11·1$ (cf. ref. 17a).

does not arise in the nitration of nitrobiphenyls. Comparisons of f_p for the four compounds show the $4'$-position to be deactivated 32,11, and 29 times by 2-, 3- and 4-nitro-groups. The data for nitrobenzene (table 9.4) show that transmission of the effect of a nitro-group through the biphenyl system rather than through a benzene ring diminishes that effect about 10^7 times.

The general phenomenon of $o:p$-substitution into the ring other than that containing a deactivating group is shown in the nitration of 4-acyl- and 4-carboxy-biphenyl,[13e] and of 4-halogenobiphenyls.[19] 4,4'-Dihalogenobiphenyls are mono-nitrated at $C_{(2)}$, whilst further nitration gives mainly the 2,3'dinitro-compound with a smaller proportion of the 2,2'-compound.[20]

Some work has been done on the nitration of polyphenyls;[17a] the case of *p*-terphenyl presents problems in connection with isomer distributions similar to those met with that of biphenyl.[21]

The series of compounds biphenyl, diphenylmethane, and fluorene is an interesting one. The following diagrams give the partial rate factors*

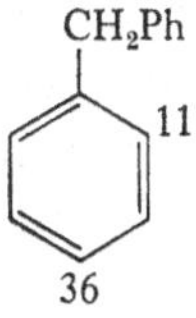

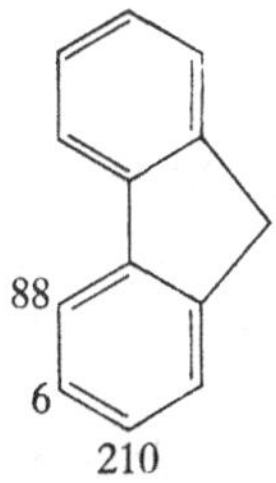

Relative rate...   19·5   101·3

for nitration at 25 °C with a solution prepared from nitric acid and acetic anhydride.[19] The benzyl group of diphenylmethane is less activating than the methyl group of toluene (table 9.1) (cf. other substituents of the type ·CH_2X, §9.1.1). The difference between biphenyl and fluorene, seen in the larger partial rate factors for $C_{(2)}$ and $C_{(4)}$ in fluorene, is presumably due chiefly to the increased conjugation of the aromatic rings in fluorene, imposed by the methylene bridge.

Biphenylene is nitrated with nitric acid in acetic acid at $C_{(2)}$, and further nitration with mixed acid gives 2,6-dinitrobiphenylene.[23] The relative rate was not determined.

10.2 NAPHTHALENE AND ITS DERIVATIVES

The nitration of naphthalene and its derivatives has been much studied, but quantitative data are still not extensive. Naphthalene itself is nitrated

---

* These partial rate factors have been recalculated from the experimental data of Dewar and Urch.[19] Their reported values for diphenylmethane are not seriously discrepant with the values now given, but this is not so for the values for fluorene. As given,[19] and copied in the literature, the values were: $f_2 = 2040$; $f_3 = 60$; $f_4 = 944$. There are consequent errors in table 8 and figs. 16 (reproduced as fig. 9.1 of this volume) and 32 of ref. 22.

Aromatic reactivity: D. Bi- and poly-cyclic compounds

mainly at $C_{(1)}$, earlier results indicating that only about 5 % of reaction occurs at $C_{(2)}$ though, as would be expected, the proportion of 2-nitration rises with temperature.[24] Solutions prepared from nitric acid and acetic anhydride gave 1- and 2-nitronaphthalene in the ratio 10:1 at 0 °C. The ratio was somewhat lower at 45 °C.[1b] For this reaction at 0 °C the rate of nitration relative to that of benzene was 400. The corresponding partial rate factors ($f_1 = 470$; $f_2 = 50$)[1e] are commented on below (§10.3); they should be compared with those ($f_1 = 212 \pm 2$; $f_2 = 11\cdot4 \pm 1\cdot2$), based on a relative rate of 149 ± 15, for nitration in acetic acid at 25 °C.[25] Recently, the ratio of 1- to 2-nitration was found to vary from about 8 to about 30, according to the conditions used[26a] (table 10.2) and from 18·6 at 25 °C to 11·2 at 100 °C for nitration in acetic acid.[25] The rate of nitration of naphthalene in 60–70 % sulphuric acid or 61 % perchloric acid is not far removed from the encounter rate, and 1- and 2-methylnaphthalene and 1-methoxynaphthalene react at the encounter rate. Below 65 % sulphuric acid the kinetics of nitration of naphthalene become erratic unless nitrous acid is removed from the solution. Clearly, nitrosation of naphthalene and subsequent oxidation of the product by nitric acid can occur (§4.3).[12] In 7·5 % aqueous sulpholan at 25 °C naphthalene reacts about 33 times faster than benzene, but 1- and 2-methyl- and 1,6-dimethyl-naphthalene react upon encounter. Under these conditions 1-methylnaphthalene, and even more readily 1,6-dimethylnaphthalene, can be nitrated via nitrosation (§4.3.4).[27]

Derivatives of naphthalene containing an activating substituent are nitrated mainly in the ring containing the substituent. When the substituent is at $C_{(1)}$ nitration occurs mainly at $C_{(4)}$, to a smaller degree at $C_{(2)}$ and also at $C_{(5)}$ and $C_{(8)}$. When the substituent is at $C_{(2)}$ nitration occurs mainly at $C_{(1)}$, and to a smaller degree at $C_{(4)}$, $C_{(5)}$, $C_{(6)}$ and $C_{(8)}$.[17b,24] The nitration of 1- and 2-methylnaphthalene under a variety of conditions has been studied, and isomer distributions and partial rate factors determined.[26a] Except for nitration with solutions prepared from nitric acid and acetic anhydride the results (table 10·2), as regards substrate selectivity, resemble those obtained by Olah and his co-workers for the nitration of toluene (§4.4.3). The meaning of these results, and even more so of those obtained similarly for the nitration of 1- and 2-methoxynaphthalene,[26b] is obscured by circumstances mentioned in the discussion of Olah's work and in the above discussion of the nitration of naphthalenes; nitration at the encounter rate, slow mixing

TABLE 10.2   *The nitration of naphthalene and its derivatives*[26a] *at 25 °C*

Compound	Reagent*	Ratio of 1- to 2-nitration
Naphthalene	A	28·5
	B	20·7
	C	21·6
	D	9·6
	E	8·6

Compound	Reagent*	Relative† rate	Isomer proportions %				
			2	3	4	5	8
1-Methylnaphthalene	A	18·4	7·4	1·2	73·5	11·1	6·8
	B	28·1	4·7	1·4	59·9	11·0	23·0
	C	2·4	7·7	1·4	63·4	14·3	13·2
	D	1·4	36·7	1·2	43·9	6·3	11·9
	E	2·2	23·3	6·1	43·0	10·1	17·5

Compound	Reagent*	Relative† rate	1	3‡	4	5	6	8
2-Methylnaphthalene	A	18·5	66·2	0·13	10·1	6·6	1·7	15·2
	B	26·3	67·5	0·14	8·1	7·0	2·3	15·0
	C	2·5	63·6	0·34	10·0	7·4	2·2	17·4
	D	2·2	56·6	0·6	18·7	7·0	3·0	14·1
	E	2·6	52·7	0·6	17·1	9·0	6·6	13·7

* A, HNO_3–$MeNO_2$; B, HNO_3–AcOH; C, HNO_3–H_2SO_4–AcOH; D, $NO_2{}^+BF_4{}^-$–sulpholan; E, HNO_3–Ac_2O.

† Rate relative to that for naphthalene as determined by competition.

‡ Calculated values.[26a]

of solutions, and nitration *via* nitrosation may be present. The significance of a M.O.-theoretical treatment of these and related results[25] is therefore uncertain.

Mixed acid nitrates the naphthylamines, presumably as their cations, at the α-positions of the ring other than that containing the substituent.[17b, 24] 1-Acetylaminonaphthalene is nitrated at the 2- and 4-positions with nitric acid in acetic acid, nitric acid alone, and nitric acid with boron trifluoride in acetic acid;[28] under these and other conditions[29] the ratio of 4- to 2-nitration (about 2·5) is remarkably constant compared with the *p:o*-ratio for the nitration of acetanilide (§5.3.4). 5-, 6- and 7-nitro groups do not greatly change the ratio of 4- to 2-nitration, but an 8-nitro group greatly increases the ratio, a result attributed to a steric effect. 2-Acetylaminonaphthalene is nitrated mainly at $C_{(1)}$, some 6- and 8-nitration also occurring. The results for the nitration of acetylamino-nitronaphthalenes show, as a whole, that reaction at β-positions only occurs when strongly activating groups are present; otherwise, the preference for substitution at an α-position dominates and a substituent merely affects the choice amongst available α-positions.[28]

The tendency towards α-substitution is also seen in the reactions of naphthalene derivatives containing de-activating substituents.[17b, 24] Thus, 1-nitronaphthalene with mixed acid gives 1,5- and 1,8-dinitronaphthalene (with some trinitronaphthalene) in the ratio of about 1:2, whilst 2-nitronaphthalene gives 1,6- and 1,7-dinitronaphthalene (with some 1,3,8-trinitronaphthalene).[30]

10.3 POLYCYCLIC HYDROCARBONS

Some information is available about the nitration of polycyclic hydro-carbons and their derivatives, but it is of no quantitative significance.[17b, 24b] The formation of a σ-complex from anthracene and nitronium ions has been mentioned (§6.2.3, §6.3).

The significance of Dewar's results for a series of polynuclear hydro-carbons, as well as for various compounds containing hetero atoms, has been discussed (§5.3.2). Though the differences are not often important, we have not in all cases been able to reproduce the values for the partial rate factors reported by these authors, by recalculation from their reported results; in table 5.3 the figures in parentheses are some examples of our recalculations.

10.4 HETEROCYCLIC COMPOUNDS

10.4.1 *2-Phenylpyridine and related compounds*

The nitration of phenylpyridines and related compounds has attracted attention for a long time, and measurements of isomer proportions have been made for several compounds of this type.[31b, 32] Nitration occurs in the phenyl ring. For 2-phenylpyridine and 2-phenylpyridine 1-oxide measurements of the dependence of rate of nitration upon acidity in 75–81 % sulphuric acid at 25 °C show that both compounds are nitrated as their cations (table 8.1). The isomer distribution did not depend significantly upon the acidity, and by comparison with the kinetic data for quinolinium (§10.4.2) the partial rate factors illustrated below were obtained.[33] They should be compared with those for the nitration of 2-nitrobiphenyl (§10.1). The protonated heterocyclic groups are much more powerfully deactivating than the nitrophenyl group. Although the oxide cation is more reactive than 2-phenylpyridinium the proportion of *m*-substitution is greater in the former. The overall similarity to the partial rate factors[33] for the benzyltrimethylammonium cation (§9.1.2) is noteworthy.

10.4.2 *Azanaphthalenes and their derivatives*

The compounds to be discussed are the quinolines, isoquinolines, cinnolines, quinazolines, quinoxalines, and phthalazines. Once again, this is a family of compounds for which much qualitative, but little quantitative information is available.

Leaving aside quinazoline, the parent heterocycles of this series and their derivatives containing weakly activating groups or deactivating groups are nitrated at the α-positions of their carbocyclic rings (except, of course, when these are blocked), and sometimes to a small extent at $C_{(6)}$.[31, 34, 35] The greater reactivity of the α-positions recalls the behaviour of the naphthalenes (§10.2).

The behaviour of quinazoline removes it from this general description; nitration in sulphuric acid gives 6-nitroquinazoline.[31a, 36] It has been suggested that this result may arise because the entity being nitrated is the hydrated quinazolinium cation,[37] but the anhydrous di-cation is evidently the dominant species in strongly acidic solutions.[38] Kinetic studies of the nitration of quinazoline in sulphuric acid were frustrated by the decomposition which occurs. Phthalazine is not easily nitrated. It is unaffected by mixtures of concentrated sulphuric acid and fuming nitric acid at 0 °C, and at higher temperatures these mixtures cause oxidation to phthalic acid, as does fuming nitric acid alone. Nitration, without oxidation, can be achieved by the use of potassium nitrate in 98% sulphuric acid, the product being exclusively 5-nitrophthalazine (79%).[39]

The preparative nitration of quinoline in mixed acid has been described several times, and has usually been carried out under unnecessarily severe conditions; good yields of 5- and 8-nitroquinoline in roughly equal amounts were obtained.[31b] In similar conditions isoquinoline is also efficiently nitrated, but 5-nitro-isoquinoline is by far the dominant product, and 8-nitro-isoquinoline is not an easily available compound.[31b] With mixed acid cinnoline gives roughly equal amounts of 5- and 8-nitrocinnoline.[40]

From the nitration of quinoline with acetic anhydride as the solvent and a mixture of metal nitrates, nitric acid or nitrogen dioxide as the nitrating agent, the main product was found to be 3-nitroquinoline with traces of the 6- and 8-isomers. In per-nitrous acid however, the 5-, 6-, 7- and 8-nitro isomers were isolated from the reaction mixture. In all cases the yield of mono-nitro-compounds was very low and up to 70 % of unchanged quinoline was recovered. 3-Nitroquinoline was thought to be formed by an addition–elimination mechanism,[41a] but has also been regarded, less probably, as arising from nitration of quinoline as the free base.[35] With dinitrogen tetroxide quinoline gives 7-nitro- and 5,7-dinitro-quinoline.[24b]

The first quantitative studies of the nitration of quinoline, isoquinoline, and cinnoline were made by Dewar and Maitlis,[41b] who measured isomer proportions and also, by competition, the relative rates of nitration of quinoline and isoquinoline (1:24·5). Subsequently, extensive kinetic studies were reported for all three of these heterocycles and their methyl quaternary derivatives (table 10.3). The usual criteria established that over the range 77–99 % sulphuric acid at 25 °C quinoline reacts as its cation (I), and the same is true for isoquinoline in 71–84 % sulphuric acid at 25 °C and 67–73 % sulphuric acid at 80 °C (§8.2; tables 8.1, 8.3). Cinnoline reacts as the 2-cinnolinium cation (IIIa) in 76–83 % sulphuric acid at 80 °C (see table 8.1). All of these cations are strongly deactivated. Approximate partial rate factors of $f_5 = 9 \cdot 0 \times 10^{-6}$ and $f_8 = 1 \cdot 0 \times 10^{-6}$ have been estimated for isoquinolinium.[42] The unprotonated nitrogen atom of the 2-cinnolinium (IIIa) and 2-methylcinnolinium (IIIb) cations causes them to react 287 and 200 more slowly than the related 2-isoquinolinium (IIa) and 2-methylisoquinolinium (IIb)

(I) (II; a, R = H   (III; a, R = H
       b, R = Me)   b, R = Me)

TABLE 10.3   *The nitration of azanaphthalenes and their derivatives**

Substance	Nitrating system, H_2SO_4/%	Temp./°C	Isomer proportions/% 3	5	6	7	8	Rate constants k_2/l mol⁻¹ s⁻¹	H_2SO_4/%	Arrhenius parameters Temp. range/°C	E/kJ mol⁻¹	$\log_{10}$ (A/l mol⁻¹ s⁻¹)	Ref.
Quinolinium†	77·45	.	0·0013	50·1	1·61	0·014	42·3	0·10 × 10⁻⁴	.		.		45
	80·05	.							.		.		44
	81·3	.						1·67 × 10⁻⁴	81·3	25–50	68·6	8·3	42
	85·50	.						89·2 × 10⁻⁴	.		.		45
	87·49	.						335 × 10⁻⁴	.		.		45
	89·50	.						503 × 10⁻⁴	.		.		45
	90·18	.						514 × 10⁻⁴	.		.		45
	97·96	.						95·3 × 10⁻⁴	97·95	25–45	62·8	9·00	45
1-Methylquinolinium	79·6	.						7·94 × 10⁻⁶	.		.		42
	81·3	.						4·44 × 10⁻⁵	81·3	25–50	75·3	8·8	42
	83·7	.						2·91 × 10⁻⁴	.		.		42
Isoquinolinium	71·3	.						8·13 × 10⁻⁷	.		.		42, 43
	74·6	.						7·05 × 10⁻⁶	.		.		42, 43
	78·1	.						1·14 × 10⁻⁴	.		.		42, 43
	79·6	.						4·08 × 10⁻³	.		.		42, 43
	81·3	.						2·32 × 10⁻³	81·3	25–50	59·0	7·7	42, 43
	83·7	.						2·13 × 10⁻²	.		.		42, 43
	67·7	80						1·85 × 10⁻⁵	.		.		43
	71·4	80						1·72 × 10⁻⁴	.		.		43
	73·0	80						4·57 × 10⁻⁴	.		.		43
2-Methylisoquinolinium	71·3	.						8·91 × 10⁻⁷	.		.		42, 43
	74·6	.						5·64 × 10⁻⁶	.		.		42, 43
	78·1	.						1·15 × 10⁻⁴	.		.		42, 43
	79·6	.						4·65 × 10⁻⁴	.		.		42
	81·3	.						2·58 × 10⁻³	81·3	25–50	62·3	8·4	42
	83·7	.						2·75 × 10⁻²	.		.		42
	64·4	80						3·55 × 10⁻⁶	.		.		43
	67·7	80						2·26 × 10⁻⁵	.		.		43
	70·9	80						1·42 × 10⁻⁴	.		.		43
	73·4	80						7·93 × 10⁻⁴	.		.		43

* At 25 °C unless otherwise stated.

† The total yield of nitroquinolines was about 94 %. An unidentified yellow compound is also formed.[44–5]

TABLE 10.3 (cont.)

Substance	Nitrating system H₂SO₄/%	Temp./°C	Isomer proportions/% 3	5	6	7	8	Rate constants k_2/l mol⁻¹ s⁻¹	H₂SO₄/%	Arrhenius parameters Temp. range/°C	E/kJ mol⁻¹	$\log_{10}$ (A/l mol⁻¹ s⁻¹)	Ref.
2-Cinnolinium	76·14	80	·	·	·	·	·	$1·18 \times 10^{-5}$	76·14	80–100	102	10·15	43
	77·0	80	·	·	·	·	·	$1·72 \times 10^{-5}$					43
	78·9	80	·	·	·	·	·	$7·58 \times 10^{-5}$	81·1	65–90	80·2	9·2	43
	81·1	80	·	·	·	·	·	$3·23 \times 10^{-4}$					43
	82·9	80	·	·	·	·	·	$1·10 \times 10^{-3}$					43
2-Methylcinnolinium	77·0	80	·	·	·	·	·	$3·81 \times 10^{-5}$					43
	79·1	80	·	·	·	·	·	$1·90 \times 10^{-4}$	81·19	60–80	80·0	8·7	43
	81·19	80	·	·	·	·	·	$7·35 \times 10^{-4}$					43
4-Hydroxyquinoline‡	80·2				81·0		19·0	$6·9 \times 10^{-3}$					44
	80·7				80·8		19·2	$1·27 \times 10^{-2}$	81·4	24–45	59·0	8·3	44
	81·4				81·4		18·6	$6·98 \times 10^{-2}$					44
	82·6				78·5		21·5	$1·78 \times 10^{-1}$					44
	83·2							$5·29 \times 10^{-1}$					44
	84·1												44
	84·4												44
	85·4												44
1-Methyl-4-quinolone‡	74·9		13·8		86·2		?	$1·65 \times 10^{-4}$					44
	77·4		4·5		95·5		?	$4·92 \times 10^{-3}$	81·4	25–45	59·0	7·9	44
	80·2						?	$4·24 \times 10^{-2}$					44
	81·4		6·0		94·0		?	$1·94 \times 10^{-2}$					44
	83·9												44
	84·1												44
	85·7												44
4-Methoxyquinoline‡	80·2				72·3		27·8	$6·30 \times 10^{-3}$					44
	81·4							$2·07 \times 10^{-2}$	81·4	25–45	59·0	8·0	44
	82·6							$7·99 \times 10^{-2}$					44
	84·2				73·9		26·1	$1·93 \times 10^{-1}$					44
	84·9												44
	85·1												44
4-Hydroxycinnoline§	81·2		0·96	0·38	58·4	0·36	39·9	$3·6 \times 10^{-5}$	81·2	25–45	84·9	10·7	44
	82·9							$1·31 \times 10^{-4}$					44
	84·4							$4·04 \times 10^{-4}$					44
	84·9							$1·03 \times 10^{-3}$					44
	85·6												44

Compound												Ref.
Quinoline 1-oxide‖	82·0	0	44·4 (C₍₄₎); 24·2 (C₍₆₎); 31·5 (C₍₈₎)				$1·12 \times 10^{-5}$ [$4·9 \times 10^{-6}$ for C$_{(6)}$] [$6·2 \times 10^{-6}$ for C$_{(6+8)}$]	82·0	0–25	93·3	12·5	53
	82·0	25	67·1 (C₍₄₎); 15·9 (C₍₆₎); 17·0 (C₍₈₎)				$2·36 \times 10^{-4}$ [$1·58 \times 10^{-4}$ for C$_{(4)}$] [$7·85 \times 10^{-5}$ for C$_{(4+7)}$]			67·8	7·8	53
1-Methoxyquinolinium‖	82·0						$1·59 \times 10^{-6}$					53
Isoquinoline 2-oxide	76·4						$1·08 \times 10^{-6}$					52
	78·7						$6·70 \times 10^{-5}$					52
	81·2						$5·00 \times 10^{-4}$					52
	83·1						$3·30 \times 10^{-3}$	83·1	25–50	61·5	8·3	52
2-Methoxyisoquinolinium	76·4						$3·08 \times 10^{-6}$					52
	78·7						$2·02 \times 10^{-5}$					52
	81·6						$2·33 \times 10^{-4}$					52
	83·1						$1·03 \times 10^{-3}$	83·1	25–50	64·9	8·4	52
Cinnoline 2-oxide†	64·4	80		8·3	48·1	43·6						53
	81·4	80		17·4	20·2	62·4						53
	90·0	80		22·4	4·7	72·9	$2·22 \times 10^{-4}$					53
2-Methoxycinnolinium	75·2	80					$1·22 \times 10^{-5}$					53
	76·5	80					$3·34 \times 10^{-5}$					53
	79·6	80					$2·52 \times 10^{-4}$					53
	81·5	80					$9·74 \times 10^{-4}$					53

† Total yields were nearly quantitative for 4-hydroxy- and 4-methoxy-quinoline, but poorer for 1-methyl-4-quinolone.

§ About 5 % of an unidentified product was formed as well as the proportions of nitro-compounds given for nitration in 84·9 % sulphuric acid.

‖ Fuller data will be found in the original paper.

Aromatic reactivity: D. Bi- and poly-cyclic compounds

cations. The isomer proportions arising from the nitration of isoquinolinium show that the hetero-group at the β-position selectively deactivates $C_{(8)}$ of the parent naphthalene skeleton, whilst the extra nitrogen atom of 2-cinnolinium then selectively deactivates $C_{(5)}$.[43]

A more detailed study of the nitration of quinolinium (1) in 80·05 % sulphuric acid at 25 °C, using isotopic dilution analysis, has shown that 3-, 5-, 6-, 7- and 8-nitroquinoline are formed (table 10.3). Combining these results with the kinetic ones, and assuming that no 2- and 4-nitration occurs, gives the partial rate factors[44] listed in table 10.4. Isoquinolinium is 14 times more reactive than quinolinium. The strong deactivation of the 3-position is in accord with an estimated[47] partial rate factor of 10^{-19} for hydrogen isotope exchange at the 3-position in the pyridinium ion. It has been estimated that the reactivity of this ion is at least 10^5 less than that of the quinolinium ion.[45] Based on this estimate, the partial rate factor for 3-nitration of the pyridinium ion would be less than 5×10^{-17}.

TABLE 10.4 *Theoretical and experimental partial rate factors for the nitration of the quinolinium ion*

| Position | Experimental* | Theoretical† | | |
		a	*b*	*c*
3	$4·57 \times 10^{-12}$	1·1	$1·3 \times 10^{-2}$	$3·2 \times 10^{-5}$
5	$1·74 \times 10^{-7}$	2·0	$6·0 \times 10^{-2}$	$5·0 \times 10^{-4}$
6	$5·58 \times 10^{-9}$	1·3	$1·3 \times 10^{-2}$	$2·0 \times 10^{-4}$
7	$4·76 \times 10^{-11}$	0·56	$7·5 \times 10^{-4}$	$6·3 \times 10^{-6}$
8	$1·47 \times 10^{-7}$	3·6	55×10^{-2}	$3·9 \times 10^{-4}$

* Derived using the relative rates of $1·695 \times 10^3$ for benzene with respect to p-dichlorobenzene,[46] and $1·084 \times 10^4$ for p-dichlorobenzene with respect to quinoline.
† See p. 214.

Numerous M.O.-theoretical calculations have been made on quinoline and quinolinium. Comparisons of the experimental results with the theoretical predictions reveals that, as expected (see §7.2), localisation energies give the best correlation. π-Electron densities are a poor criterion of reactivity; in electrophilic substitution the most reactive sites for both the quinolinium ion and the neutral molecule are predicted to be the 3-, 6- and 8-positions.[48]

It is pertinent here to consider some of the results obtained by Greenwood and McWeeny[49] using both q_r and $\pi_{r,r}$ as criteria of reactivity (§7.2.2). They have calculated for quinoline the exact charges q'_r in the polarised molecules when $\delta\alpha_r$ due to the attacking ion takes the values $0.5\beta_0$ and $1.5\beta_0$ respectively for the atom positions $r = 3, 5, 6$ and 8. Their results are shown in table 10.5.

TABLE 10.5   *Charges q_r' and energy changes ΔE*
in the polarised quinoline molecule

Position	q_r'	$\Delta E(-\beta_0)$	q_r'	$\Delta E(-\beta_0)$
3	1·206	0·277	1·522	0·964
5	1·205	0·274	1·544	0·968
6	1·201	0·276	1·520	0·961
8	1·226	0·280	1·555	0·982
$\delta\alpha_r$	$0.5\beta_0$		$1.5\beta_0$	

It is clear that as the value given to $\delta\alpha_r$, the polarization parameter, increases, the charge builds up more rapidly at the site of attack when this coincides with the 5- or 8-positions than when it coincides with position 3. The corresponding π-electron energy changes show that as $\delta\alpha_r$ increases, positions 8 and 5 overtake position 3 as the active sites. The discrepancies between the theoretical predictions and experimental results begin to disappear when $\delta\alpha_r$ is given the same value as $\delta\alpha_u$, the change in coulomb integral at position u occupied by a hetero-atom. Thus if the polarization effects due to the attacking ion outweigh the effects due to the nitrogen atom positions 5 and 8 are correctly predicted to be the most reactive. Clearly when both q_r and $\pi_{r,r}$ are considered a more realistic picture is obtained.

Several workers have calculated localisation energies for the electrophilic substitution of quinoline or quinolinium.[41b, 48d,f, 50a] These generally succeed in predicting that the α-positions of the carbocyclic ring will be the most reactive, followed by $C_{(6)}$. To the same extent, calculations for isoquinolinium predict orientation correctly.[50b] Dewar's method of calculating approximate localisation energies (§7.2.3) has been applied to the six-membered nitrogen heterocyclic compounds;[41b] it performs well in giving a rough qualitative account of the orientation to be expected in nitration of the cations. It is not satisfactory in detail, as can be seen from the predicted order of reactivity of positions in the quinolinium ion [(8) > 5 > 6 = 3 > 7 ≫ 4 ≫ (2)]. As regards quinazoline, it is not certain that the model is relevant (see above). The absolute values of the approximate localisation energies have no significance, depending as they do on such assumptions as that the nitrating agent in sulphuric acid is a more reactive agent than that operating in solutions prepared from nitric acid and acetic anhydride, and so requires a different value of β_R (§7.2.3).

Localisation energies do not perform well in revealing the strong deactivation of the heterocyclic cations. In table 10.4 there are shown, beside the experimental values of the partial rate factors for the nitration of the quinolinium ion, two sets of values derived from localisation energies:[42,44] those in column *a* from the localisation energies of Brown and Harcourt,[50a] and those in column *b* from the localisation energies of Dewar and Maitlis.[41b] Column *c* gives the results of an attempt[42,44] to take into account the likelihood that constant entropy of activation cannot be assumed since a reaction between two positive ions is being compared with a reaction between an ion and a neutral molecule. Such 'corrections' cannot be taken very seriously because of the assumptions which have to be made in applying them, and also because of the possibility that changes in $\Delta S^{\ddagger}$ might be compensated for by changes in $\Delta H^{\ddagger}$.

Little is known quantitatively about substituent effects in the nitration of derivatives of azanaphthalenes. In preparative experiments 4-hydroxy-quinoline, -cinnoline, and -quinazoline give the 6- and 8-nitro compounds, but with nitric acid alone 4-hydroxyquinoline and 2,4-di-hydroxyquinoline react at $C_{(3)}$.[31] With nitric acid, 4-hydroxycinnoline still gives mainly 4-hydroxy-6-nitrocinnoline, but some of the 3-nitro compound can also be isolated.[31,51] The change of orientation with reagent could be due to a change to free-base nitration in the more weakly acidic medium, or to the occurrence in nitric acid of nitration *via* nitrosation.[31]

Recently kinetic data have become available for the nitration in sulphuric acid of some of these hydroxy compounds (table 10.3).[44] For 4-hydroxyquinoline and 4-methoxyquinoline the results verify the early conclusions[31] regarding the nature of the substrate being nitrated in sulphuric acid. Plots of $\log_{10} k_2$ against $-(H_R + \log_{10} a_{H_2O})$ for these compounds and for 1-methyl-4-quinolone have slopes of 1·0, 1·0 and 0·97 at 25 °C respectively, in accord with nitration *via* the majority species (§8.2) which is in each case the corresponding cation of the type (IV). At a given acidity the similarity of the observed second-order rate constants for the nitrations of the quinolones and 4-methoxy-quinoline at 25 °C supports the view that similarly constructed cations are involved. Application of the encounter criterion eliminates the possibilities of a free base nitration in the case of 4-methoxyquinoline, but for the other two compounds this criterion does not assist the choice between nitration *via* the free base and the conjugate acid.

(IV)

In the cases of 4-hydroxyquinoline and 4-methoxyquinoline the predominance of 6-nitration supports the evidence presented above that nitration in sulphuric acid proceeds *via* the cations. For both these compounds reaction *via* the neutral molecules would be expected to occur to a considerable extent at $C_{(3)}$ as a result of the directing properties of the hydroxyl and methoxyl groups.

The case of 1-methyl-4-quinolone is puzzling. The large proportion of the 3-nitro isomer formed in the nitration (table 10.3; cf. 4-hydroxyquinoline) might be a result of nitration *via* the free base but this is not substantiated by the acidity dependence of the rate of nitration or by the Arrhenius parameters. From 1-methyl-4-quinolone the total yield of nitro-compounds was not high (table 10.3).

With 4-hydroxycinnoline the situation is not clear. The slope of $\log_{10} k_2$ *v.* $-(H_R + \log_{10} a_{H_2O})$ is 0·84, a lower value than is normally observed for compounds reacting *via* the majority species (table 8.1). In this instance use of the encounter rate criterion (§8.2.3) is not helpful; the calculated encounter rate constant turns out to be much larger than the observed rate constant.[44] Examination of the products of nitration of 4-hydroxycinnoline in 84·9 % sulphuric acid at 25 °C (table 10.3) reveals a surprisingly large proportion of 4-hydroxy-3-nitrocinnoline. In view of the obvious deactivation of this position with respect to electrophilic substitution by the adjacent nitrogen atom and the fact that no 4-hydroxy-3-nitroquinoline could be detected in the nitration of 4-hydroxy-quinoline under similar conditions, nitration of the 4-hydroxy-cinnolinium cation would not be expected to give any of the 3-nitro isomer. It seems likely therefore that this isomer is formed by nitration of the neutral molecule.

As regards the nitration of these compounds in nitric acid alone, mentioned above, the situation is still not clear. However, with 4-hydroxyquinoline the presence of urea made no difference to the formation of 4-hydroxy-3-nitroquinoline, and the available results are not inconsistent with the view that in nitric acid it is the free bases which react.[44]

Partial rate factors for the nitration of 4-hydroxyquinoline and its derivatives are given in table 10.6. Comparison with the values for quinolinium (table 10.4) show that the introduction of a 4-hydroxy or a 4-methoxy group into the latter activates the 6-position by factors of $3·3 \times 10^3$ and $1·6 \times 10^3$, respectively, and the 8-position by factors of $29·5$ and 23, respectively.[44] What has been said above makes the significance of partial rate factors which may be calculated for 4-hydroxy-cinnoline uncertain.

TABLE 10.6 *Partial rate factors* for the nitration of 4-hydroxy-quinoline and related compounds in 80·05% sulphuric acid at 25 °C*

Compound	Position				
	3	5	6	7	8
4-Hydroxyquinoline	.	.	$1·84 \times 10^{-5}$	.	$4·34 \times 10^{-6}$
4-Methoxyquinoline	.	.	$8·84 \times 10^{-6}$	.	$3·34 \times 10^{-6}$
1-Methyl-4-quinolone	$3·85 \times 10^{-7}$	.	$8·14 \times 10^{-6}$	.	.

* Calculated using data for benzene and *p*-dichlorobenzene.

Among the *N*-oxides of this series of compounds isoquinoline 2-oxide shows the simplest behaviour on nitration. The acidity dependence of the rate of nitration (table 8.1), and comparison with the 2-methoxyiso-quinolinium cation (v, $R = $ Me) (table 10.3) show the oxide to be nitrated as its conjugate acid (v, $R = $ H) in 76–83% sulphuric acid. The

(V) (VI) (VII)

isomer distribution is not accurately known but approximate partial rate factors ($f_5 = 5·4 \times 10^{-6}$; $f_8 = 6·0 \times 10^{-7}$) show the strong deactivation caused by the protonated oxide function. The cations isoquinolinium,

2-methylisoquinolinium, 2-hydroxyisoquinolinium, and 2-methoxy-isoquinolinium show relative reactivities in 76·3 % sulphuric acid at 25 °C of 1:1:0·47:0·13.[52]

Comparison of the behaviour of cinnoline 2-oxide (VI, $R = \overline{O}$) with that of 2-methoxycinnolinium (VI, $R = $ OMe) suggests that at high acidities the former is nitrated as its conjugate acid (VI, $R = $ OH), but that as the acidity is lowered the free base becomes active. At high acidities 5- and 8-nitration are dominant, but as the acidity is lowered 6-nitration becomes increasingly important. The 5- and 8-nitro compounds are probably formed mainly or wholly by nitration of the conjugate acid, and the 6-nitro compound wholly or mainly from the free base.[53]

Ochiai and Okamoto[54] showed that nitration of quinoline 1-oxide in sulphuric acid at 0 °C gave 5- and 8-nitroquinoline 1-oxides with a trace of the 4-isomer, but that at 60–100 °C 4-nitration became overwhelmingly dominant. The orientation depends not only upon temperature but also upon acidity, and kinetic studies (table 8.4; table 10.3) show that two processes are occurring: the nitration of the free base (VII, $R = \overline{O}$ at $C_{(4)}$, favoured by low acidities and high temperatures, and the nitration of the cation (VII, $R = $ OH), favoured by high acidities and low temperatures.[53]

As with cases mentioned earlier, Hückel M.O. theory performs satisfactorily in predicting the orientation of nitration in these oxides, but again fails to reproduce their strong deactivation.[55]

10.4.3 *Benzimidazoles*

Benzimidazole (VIII) and indazole (IX) differ from the azanaphthalenes in being nitrated at the β-position $[C_{(5)}]$ of the carbocyclic ring.[17a, 31b]

(VIII)   (IX)

The same orientation is found in the nitration of 2-methylbenzimidazole, whilst 5-nitro- and 2-methyl-5-nitrobenzimidazole are further nitrated at $C_{(6)}$. The acidity dependence of the rate of nitration of

217

Aromatic reactivity: D. Bi- and poly-cyclic compounds

2-methyl- and 1,2-dimethyl benzimidazole, and comparison with that for 1,2,3-trimethylbenzimidazolium, show the two former compounds to be nitrated as their cations in 70–76% sulphuric acid.[56] Hückel M.O. calculations fail badly with benzimidazole. Localisation energies for the free base and the cation indicated $C_{(4)}$ to be the most reactive position towards electrophilic attack,[55,57] and led to the false conclusion that substitution involved the free base, the orientation being controlled by charge[57] densities.

2-Phenylbenzimidazole is nitrated first at the 5-position with mixed acid, and subsequent reaction produces 5-nitro-2-(4-nitrophenyl)- and 5-nitro-2-(3-nitrophenyl)-benzimidazole. 2-Phenyl-, 2-(4-nitrophenyl)- and 5-nitro-2-phenyl-benzimidazole are nitrated as their conjugate acids.[56b]

REFERENCES

1. (a) Bavin, P. M. G. & Dewar, M. J. S. (1956). *J. chem. Soc.* p. 164.
   (b) Dewar, M. J. S. & Mole, T. (1956). *J. chem. Soc.* p. 1441.
   (c) Dewar, M. J. S. & Warford, E. W. T. (1956). *J. chem. Soc.* p. 3570.
   (d) Dewar, M. J. S., Mole, T., Urch, D. S. & Warford, E. W. T. (1956). *J. chem. Soc.* p. 3572.
   (e) Dewar, M. J. S., Mole, T. & Warford, E. W. T. (1956). *J. chem. Soc.* p. 3576.
   (f) Dewar, M. J. S., Mole, T. & Warford, E. W. T. (1956). *J. chem. Soc.* p. 3581.
   (g) Dewar, M. J. S. & Urch, D. S. (1958). *J. chem. Soc.* p. 3079.
2. Simamura, O. & Mizuno, Y. (1957). *Bull. chem. Soc. Japan.* **30**, 196. Mizuno, Y. & Simamura, O. (1958). *J. chem. Soc.* p. 3875.
3. Hayashi, E., Inana, K. & Ishikawa, T. (1959). *J. pharm. Soc. Japan* **79**, 972.
4. Billing, C. J. & Norman, R. O. C. (1961). *J. chem. Soc.* p. 3885.
5. Bell, F., Kenyon, J. & Robinson, P. H. (1926). *J. chem. Soc.* p. 1239.
6. Jenkins, R. L., McCullough, R. & Booth, C. F. (1930). *Ind. Engng Chem.* **22**, 31.
7. Shorygin, P. P., Topchiev, A. V. & Anan'ina, V. A. (1938). *Zh. obshch. Khim.* **8**, 981.
8. de la Mare, P. B. D. & Hassan, M. (1957). *J. chem. Soc.* p. 3004.
9. Buck, K. R. & Thompson, R. S. (1962). *Chemy Ind.* p. 882.
10. Taylor, R. (1966). *J. chem. Soc. B*, p. 727.
11. Taylor, R. (1966). *Tetrahedron Lett.* p. 6093.
12. Coombes, R. G., Moodie, R. B. & Schofield, K. (1968). *J. chem. Soc. B*, p. 800.
13. (a) Grieve, W. S. M. & Hey, D. H. (1932). *J. chem. Soc.* p. 1888.
    (b) Grieve, W. S. M. & Hey, D. H. (1932). p. 2245.
    (c) Hey, D. H. (1932). *J. chem. Soc.* p. 2637.

(d) Hey, D. H. & Jackson, E. R. B. (1934). *J. chem. Soc.* p. 645.

(e) Grieve, W. S. M. & Hey, D. H. (1933). *J. chem. Soc.* p. 968.

14. Bell, F. (1928). *J. chem. Soc.* p. 2770.

15. Scarborough, H. A. & Waters, W. A. (1927). *J. chem. Soc.* (a) p. 89; (b) p. 1133.

16. Blakey, W. & Scarborough, H. A. (1927). *J. chem. Soc.* p. 3000.

17. de la Mare, P. B. D. & Ridd, J. H. (1959). *Aromatic Substitution: Nitration and Halogenation.* London: Butterworths. (a) ch. 12; (b) ch. 13; (c) ch. 15.

18. Gull, H. C. & Turner, E. E. (1929). *J. chem. Soc.* p. 491.

19. van Hove, T. (1922). *Bull. Acad. Belg. Cl. Sci.* (5) **8**, 505.

    Case, F. H. (1942). *J. Am. chem. Soc.* **64**, 1848.

    Case, F. H. & Schock, R. U. (1943). *J. Am. chem. Soc.* **65**, 2086.

    Mascarelli, L., Gatti, D. & Longo, B. (1933). *Gazz. chim. ital.* **63**, 654.

20. Dennett, H. G. & Turner, E. E. (1926). *J. chem. Soc.* p. 476.

    Le Fèvre, R. J. W. & Turner, E. E. (1926). *J. chem. Soc.* p. 2041.

    Le Fèvre, R. J. W., Moir, D. D. & Turner, E. E. (1927). *J. chem. Soc.* p. 2330.

    Le Fèvre, R. J. W. & Turner, E. E. (1930). *J. chem. Soc.* p. 1158.

    Marler, E. E. J. & Turner, E. E. (1931). *J. chem. Soc.* 1359.

    Shaw, F. R. & Turner, E. E. (1932). *J. chem. Soc.* pp. 285, 509.

21. Gray, C. W. & Lewis, D. (1961). *J. chem. Soc.* p. 5156.

22. Stock, L. M. & Brown, H. C. (1963). *Adv. phys. org. Chem.* **1**, 35.

23. Baker, W., Barton, J. W. & McOmie, J. F. W. (1958). *J. chem Soc.* p. 2666.

24. (a) Donaldson, N. (1958). *The Chemistry and Technology of Naphthalene Compounds.* London: Arnold.

    (b) Topchiev, A. V. (1959). *Nitration of Hydrocarbons and other Organic Compounds.* London: Pergamon.

25. Clark, D. J. & Fairweather, D. J. (1969). *Tetrahedron*, p. 5525.

26. Alcorn, P. G. E. & Wells, P. R. (1965). *Aust. J. Chem.* (a) p. 1377; (b) p. 1391.

27. Hoggett, J. G., Moodie, R. B. & Schofield, K. (1969). *J. chem. Soc.* B, p. 1.

28. Ward, E. R. & Wells, P. R. (1961). *J. chem. Soc.* p. 4859.

29. Hodgson, H. H. & Walker, J. (1933). *J. chem. Soc.* p. 1205.

30. Hodgson, H. H. & Walker, J. (1933). *J. chem. Soc.* p. 1346.

    Ward, E. R. & Hawkins, J. G. (1954). *J. chem. Soc.* p. 2975.

31. (a) Schofield, K. & Swain, T. (1949). *J. chem. Soc.* p. 1367.

    (b) Schofield, K. (1950). *Q. Rev. chem. Soc.* **4**, 382.

32. Schofield, K. (1967). *Hetero-aromatic Nitrogen Compounds: Pyrroles and Pyridines.* London: Butterworths.

33. Katritzky, A. R. & Kingsland, M. (1968). *J. chem. Soc.* B, p. 862.

34. Dewar, M. J. S. & Maitlis, P. M. (1957). *J. chem. Soc.* p. 2518.

35. Ridd, J. H. (1963). In *Physical Methods in Heterocyclic Chemistry* (ed. A. R. Katritzky). New York: Academic Press.

36. Elderfield, R. C., Williamson, T. A., Gensler, W. J. & Kremer, C. B. (1947). *J. org. Chem.* **12**, 405.

37. Schofield, K. (1957). *Chemy Ind.* p. 1068.

38. Armarego, W. L. F. (1967). *Quinazolines.* New York: Interscience.

39. Parsons, P. G. & Rodda, H. J. Personal communication.

References

40. Morley, J. S. (1951). *J. chem. Soc.* p. 1971.
    Alford, E. J. and Schofield, K. (1953). *J. chem. Soc.* p. 609.
41. Dewar, M. J. S. & Maitlis, P. M. (1957). *J. chem. Soc.* (a) p. 944; (b) p. 2521.
42. Moodie, R. B., Schofield, K. & Williamson, M. J. (1964). In *Nitro-Compounds*, p. 89. Proceedings of International Symposium, Warsaw. London: Pergamon Press,
43. Moodie, R. B., Qureshi, E. A., Schofield, K. & Gleghorn, J. T. (1968). *J. chem. Soc.* B, p. 312.
44. Penton, J. R. (1969). Ph.D. thesis, University of Exeter.
45. Austin, M. W. & Ridd, J. H. (1963). *J. chem. Soc.* p. 4204.
46. Coombes, R. G., Crout, D. H. G., Hoggett, J. G., Moodie, R. B. & Schofield, K. (1970). *J. chem. Soc.* B, p. 347.
47. Katritzky, A. R. & Ridgewell, B. J. (1963). *J. chem. Soc.* p. 3753.
48. (a) Coulson, C. A. & Longuet-Higgins, H. C. (1947). *Proc. Roy. Soc. Lond.* A **192**, 16.
    (b) Coulson, C. A. & Longuet-Higgins, H. C. (1947). *Trans. Faraday Soc.* **43**, 87.
    (c) Coulson, C. A. & Longuet-Higgins, H. C. (1949). *J. chem. Soc.* p. 971.
    (d) Sandorfy, C. & Ivan, P. (1950). *Bull. Soc. chim. Fr.* **17**, 131.
    (e) Coppens, G. & Nasielski, J. (1962). *Tetrahedron* **18**, 507.
    (f) Zahradník, R. & Párkányi, C. (1965). *Colln Czech. chem. Commun.* **30**, 355.
    (g) Flurry, R. L., Stout, E. W. & Bell, J. J. (1957). *Theor. chim. Acta* **8**, 203.
    (h) Nishimoto, K. (1968). *Theor. chim. Acta* **10**, 65.
    (i) Nishimoto, K. & Forster, L. S. (1966). *Theor. chim. Acta* **4**, 155.
    (j) Adam, W. & Grimison, A. (1965). *Tetrahedron* **21**, 3417.
49. Greenwood, H. H. & McWeeny, R. (1966). In *Advances in Physical Organic Chemistry*, (ed. V. Gold), vol. 4. London: Academic Press.
50. Brown, R. D. & Harcourt, R. D. (a) (1959). *J. chem. Soc.* p. 3451; (b) (1960). *Tetrahedron* **8**, 23.
51. Baumgarten, H. E. (1955). *J. Am. chem. Soc.* **77**, 5109.
52. Gleghorn, J., Moodie, R. B., Schofield, K. & Williamson, M. J. (1966). *J. chem. Soc.* B, p. 870.
53. Gleghorn, J. T., Moodie, R. B., Qureshi, E. A. & Schofield, K. (1968). *J. chem. Soc.* B, p. 316.
54. Ochiai, E. & Okamoto, T. (1950). *J. pharm. Soc. Japan* **70**, 22.
55. Gleghorn, J. T. (1967). Ph.D. thesis, University of Exeter.
56. Štěrba, V. & Arient, J. (a) with Navrátil, F. (1966). *Colln Czech. chem. Commun.* **31**, 113; (b) with Šlosar, J. (1966). *Colln Czech. chem. Commun.* **31**, 1093.
57. Brown, R. D. & Heffernan, M. L. (1956). *J. chem. Soc.* p. 4288.

Appendix

We describe here work which has appeared since our text was completed. Topics are discussed under the numbers of the sections of the main text to which they are relevant.

4.3.2

Ridd[1] has reinterpreted the results concerning the anticatalysis of the first-order nitration of nitrobenzene in pure and in partly aqueous nitric acid brought about by the addition of dinitrogen tetroxide. In these media this solute is almost fully ionised to nitrosonium ion and nitrate ion. The latter is responsible for the anticatalysis, because it reduces the concentration of nitronium ion formed in the following equilibrium:

$$2HNO_3 = NO_2^+ + NO_3^- + H_2O.$$

Considering first pure nitric acid as the solvent, if the concentrations of nitronium ion in the absence and presence of a stoichiometric concentration x of dinitrogen tetroxide are y_0 and y respectively, these will also represent the concentrations of water in the two solutions, and the concentrations of nitrate ion will be y_0 and $x+y$ respectively. The equilibrium law, assuming that the variation of activity coefficients is negligible, then requires that:

$$y^2(x+y) = y_0^3.$$

In partly aqueous nitric acid, the concentration of water is constant, and the corresponding equation is:

$$y(x+y) = y_0^2.$$

Since the first-order rate constant for nitration is proportional to y, the equilibrium concentration of nitronium ion, the above equations show the way in which the rate constant will vary with x, the stoichiometric concentration of dinitrogen tetroxide, in the two media. An adequate fit between theory and experiment was thus obtained. A significant feature of this analysis is that the weak anticatalysis in pure nitric acid, and the substantially stronger anticatalysis in partly aqueous nitric acid, do not require separate interpretations, as have been given for the similar observations concerning nitration in organic solvents.

5.3 and 9.3

Cooksey, Morgan and Morrey[2] have studied the nitration of pyrrole in the presence of acetic anhydride. They conclude that acetyl nitrate is the effective reagent, because only conditions under which this compound was present lead to the formation of nitropyrroles. Such evidence of course does not preclude the operation of some nitrating agent formed from acetyl nitrate. For a variety of conditions, the ratio of 2-nitropyrrole to 3-nitropyrrole in the product was *c.* 4. The effects of varying the solvent and the temperature were investigated, and yields as high as 70% were obtained under certain circumstances. Competition

Appendix

experiments were used in an attempt to establish the reactivity of pyrrole relative to that of benzene, by stepwise comparison with compounds of intermediate reactivity. The following values of reactivities relative to benzene were reported: benzene, 1; toluene, 27; p-xylene, 92; thiophen, 183; mesitylene, 1750; pyrrole, 53000; diphenylamine, 738000. The value for diphenylamine is in excellent agreement with that found by Dewar (table 5.3), but the significance of the comparison of the most and least reactive compounds listed is again open to question, because it is unlikely that they share a common mechanism (§5.3.3). The value of the reactivity of toluene relative to that of benzene is in agreement with previous work, but neither the value for toluene nor that for mesitylene agree with the values reported below.

More information has appeared concerning the nature of the side reactions, such as acetoxylation, which occur when certain methylated aromatic hydrocarbons are treated with mixtures prepared from nitric acid and acetic anhydride. Blackstock, Fischer, Richards, Vaughan and Wright[3] have provided excellent evidence in support of a suggested (§5.3.5) addition-elimination route towards 3,4-dimethylphenyl acetate in the reaction of o-xylene. Two intermediates were isolated, both of which gave rise to 3,4-dimethylphenyl acetate in aqueous acidic media and when subjected to vapour phase chromatography. One was positively identified, by ultraviolet, infra-red, n.m.r., and mass spectrometric studies, as the compound (I). The other was less stable and less well identified, but could be (II).

(I)   (II)

Davies and Warren[4] found that when 1,4-dimethylnaphthalene was treated with nitric acid in acetic anhydride, and the mixture was quenched after 24 hr, a pale yellow solid with an ultraviolet spectrum similar to that of α-nitronaphthalene was produced. However, if the mixture was allowed to stand for 5 days, the product was 1-methyl-4 nitromethylnaphthalene, in agreement with earlier findings.[5] Davies and Warren suggested that the intermediate was 1,4-dimethyl-5 nitronaphthalene, which underwent acid catalysed rearrangement to the final product. Robinson[6] pointed out that this is improbable, and suggested an alternative structure (IV) for the intermediate, together with a scheme for its formation from an adduct (III) (analogous to I above) and its subsequent decomposition to the observed product.

Thompson[7] has recently obtained clear evidence, from n.m.r. studies of the reaction mixture, that the observed product (VI) is *not* produced *via* 1,4-dimethyl-5-nitronaphthalene.

Hartshorn and Thompson[8] have also found evidence for adduct formation with o-xylene, n.m.r. investigations of the reaction solution revealing peaks in

222

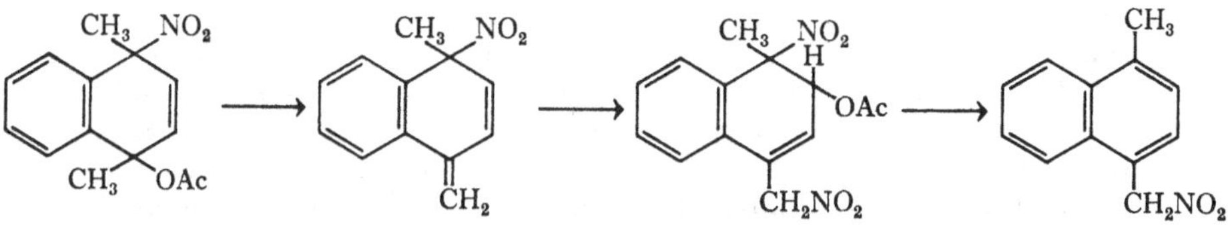

(III) (IV) (V) (VI)

the region $\tau = 3\cdot8$ to $\tau = 4\cdot5$, upfield of the peaks due to the aromatic protons. Anisole and mesitylene on the other hand, showed no such peaks, and the indications were that only nuclear nitration occurred, in agreement with the results of isolation procedures. The same workers made a more extensive study of the reaction of p-xylene. Acetyl nitrate ($0\cdot1$ mol) and p-xylene ($0\cdot2$ mol) in acetic anhydride at $0\,°C$ gave a solution which after 24 hr showed n.m.r. peaks at $\tau = 2\cdot3$ (s) and $\tau = 2\cdot77$ (s) (nitro-p-xylene), at $\tau = 3\cdot0$ (s) (p-xylene), at $\tau = 3\cdot85$ and $\tau = 3\cdot97$ (believed to be due to an adduct), and at $\tau = 4\cdot6$ (s) (benzylic CH_2). After heating the mixture, the peaks at $\tau = 3\cdot85$ and $\tau = 3\cdot97$ had disappeared, and a new peak had appeared at $\tau = 3\cdot17$ (2,5-dimethylphenyl acetate). The following products were isolated and identified: 1,4-dimethyl-2,5-dinitrobenzene, p-tolylnitromethane p-methylbenzyl acetate and 2,5-dimethylphenyl acetate. The formation of the dinitro compound is unusual and has not yet been explained, but the other products could arise *via* a Wheland intermediate formed by electrophilic attack on a methylated ring position, as outlined in the speculative scheme below.

(VII)

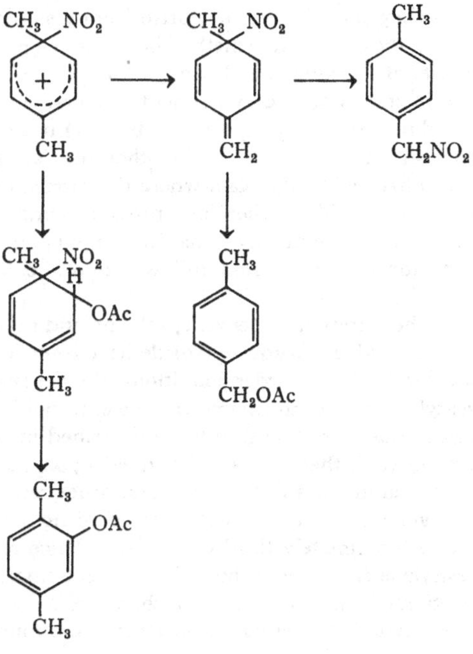

Appendix

Thompson[7] points out that there is no evidence that adducts give other than acetates on thermolysis. The exocyclic methylene intermediate (IV) postulated by Robinson could arise by proton abstraction from a Wheland intermediate analogous to (VII) above, rather than from the adduct (III). Similarly its decomposition does not necessarily require the intermediacy of the adduct (V). The fact that 1-methyl-4-nitromethylnaphthalene is the product even when the nitrating medium is nitric acid and nitromethane[5] would then require no separate explanation.

Tetralin shows evidence in n.m.r. spectroscopy, similar to that mentioned above, for the formation of one or more addition complexes. Tetralin (like indan) is known to undergo acetoxylation.[9]

It is noteworthy that the compounds which have been shown to undergo extensive acetoxylation or side-chain nitration, viz. those discussed above and hemimellitene and pseudocumene (table 5.4), are all substances which have an alkylated ring position activated towards electrophilic attack by other substituents.

Smaller amounts of these side reactions occur even with other hydrocarbons. Hartshorn et al.[10] showed that c. 2% of p-tolyl acetate, and c. 1–2% of phenyl-nitromethane, were formed when toluene was treated with a mixture prepared from nitric acid and acetic anhydride. The former product has previously been observed but the latter has not. Failure to detect either or both of these side products could account for the discrepancy between previously reported values of the reactivity of toluene relative to that of benzene towards nitration in this medium (table 9.1), and the latest value[9] of 50 ($\pm$ 6) and 38 ($\pm$ 5) determined by the kinetic and by the competition methods respectively. The latter figure takes into account ring nitro-isomers only. If the two side products are included in the estimation (as they should be if they arise from a similar rate-determining step) the figure from the competition method becomes 40 ($\pm$ 5) and the difference between the results of the kinetic and competition studies, whilst disturbingly large, comes within the combined experimental errors. Both figures are much larger than the values previously reported (table 9.1) not only for nitration in acetic anhydride but also for nitration in other media, such as nitric acid-nitromethane and nitric acid-sulpholan, where the nitronium ion is believed to be the effective reagent. They therefore provide some evidence (although indirect) that a simple nitronium ion mechanism is not operative in this medium. Further evidence for this conclusion follows from the kinetic studies now reported.

The kinetics of the nitration of benzene, toluene and mesitylene in mixtures prepared from nitric acid and acetic anhydride have been studied by Hartshorn and Thompson.[8] Under zeroth order conditions, the dependence of the rate of nitration of mesitylene on the stoichiometric concentrations of nitric acid, acetic acid and lithium nitrate were found to be as described in section 5.3.5. When the conditions were such that the rate depended upon the first power of the concentration of the aromatic substrate, the first order rate constant was found to vary with the stoichiometric concentration of nitric acid as shown on the graph below. An approximately third order dependence on this quantity was found with mesitylene and toluene, but with benzene, increasing the stoichiometric concentration of nitric acid caused a change to an approximately second order dependence. Relative reactivities, however, were found to be insensitive to the conditions, which justified the deduction of the relative reactivity for toluene mentioned above. In the same way the reactivity of mesitylene relative to that of benzene was found to be *c.* 5000. The effect of added acetic acid was complicated. With benzene in solutions containing high stoichiometric concentrations of nitric acid, acetic acid had little effect, but at lower concentrations of nitric acid, acetic acid both catalysed the reaction and reduced the order with respect to the stoichiometric concentration of nitric acid.

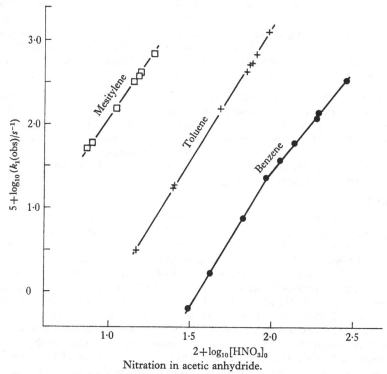

Nitration in acetic anhydride.

It has not been found possible to reconcile all these observations with a simple kinetic scheme. A major difficulty is that whilst the *stoichiometric* concentrations of nitric acid and of acetic acid can be varied independently, the actual concentrations of these species cannot, because of the existence of the equilibrium:

$$HNO_3 + Ac_2O = AcONO_2 + AcOH.$$

Another difficulty is that the extent to which hydrogen bonded association and ion-pairing influence the observed kinetics has yet to be determined. However the high order of the reaction in the stoichiometric concentration of nitric acid would seem to preclude a transition state composed only of a nitronium ion and an aromatic molecule.

6.5

Banthorpe[11] has probed the evidence that has been adduced for the intermediacy of π complexes in organic reactions. In this excellent review, the useful distinc-

Appendix

tion is drawn between molecular, or charge transfer, complexes on the one hand, and π complexes as described by Dewar on the other. With regard to aromatic substitutions, the conclusion is drawn that 'no physical evidence has been obtained, despite extensive search, that requires the postulation or accumulation of π complexes during these reactions'.

7.2.1 and 9.1.2

Katritzky and Topsom[12] have reviewed the information available, largely from infrared and n.m.r. studies, concerning the distortion of the π-electron system in the benzene ring brought about in the ground state by substituents. Of particular interest is the observation that both n.m.r. studies (of m-^{19}F and p-^{19}F chemical shifts) and infrared investigations (of the intensities of bands due to certain skeletal vibrations) suggest that the value of Taft's σ_R^0 constant for certain positively charged substituents, including $\overset{+}{N}D_3$ and $\overset{+}{N}Me_3$, is negative, which in turn implies that these groups act as π-electron *donors*. Such a character is the opposite of what would be expected from the π-inductive effect. The origin of the effect is obscure, but since it leads to increased electron density at the *para*-position it may contribute to the understanding of the observation of extensive *para*-substitution in the nitration of the anilinium ion and its N-methylated derivatives (table 9.3).

9.1.2

The substituent effects of positive poles, in particular of the trimethylammonio group, have recently been re-discussed by Ingold[13] who recalls his own earlier statement that the field effect, 'if it were strong enough, could reverse, as between *meta*- and *para*-positions, an orientation determined by the inductive effect'. His earlier conclusion was 'and still is', that 'it is not strong enough'. The consequences of the inductive effect, which 'enters at the α-carbon atom', are represented by the first of the two following diagrams, and the consequences of the field effect, which 'enters mainly... at the *ortho*-carbon atoms', by the second. The two effects combine to deactivate *ortho*-positions, but produce reversed orientations at the *meta*- and *para*-positions. The results for the phenyl trimethylammonium ion are ascribed to an inductive effect and a field effect; though the latter is strong, it is weaker than the former. The strong field effect arises because a solvent molecule cannot interpose itself between the pole and the ortho-position at which it exerts its major influence. The case is compared in these terms with that of the benzyl trimethyl ammonium ion.

Ingold introduces the terms 'substrate field effect' and 'reagent field effect' to describe those aspects of the direct field effect numbered (2) and (3) in §9.1.2. His description of the substituent effect of the trimethylammonio group is thus given substantially in terms of the substrate field effect and the π-inductive effect, i.e. it is an isolated molecule description. The reagent field effect is seen to be significant in nitration and to produce qualitatively the same result as the substrate field effect, but at this stage the two effects are not separable in aromatic substitution.

The first three of a series of papers by Ridd and co-workers on 'Inductive and Field effects in Aromatic Substitution' have appeared.[14] Results of studies of the nitration of 4-phenylpyridine and of 4-benzylpyridine in aqueous sulphuric acid were reported[14a] and use of the usual criteria (para 8.2) showed that in each case the conjugate acid was the species undergoing nitration. The values of $\log(f_m/f_m^0)$, where f_m^0 refers to the corresponding homocyclic compound (biphenyl or diphenylmethane) when plotted against r, the distance between the positive charges in the transition state (see p. 173) fell near the line drawn through the points for the compounds $Ph(CH_2)_n\overset{+}{N}Me_3$; the deactivating effect of the protonated pyridyl group can thus most readily be understood as the field effect of a pole.

These and other studies of the relative substituent effects of X and CH_2X in nitration were considered[14c] in terms of the transmission factor α of the methylene group. To avoid complications from conjugative interactions, attention was focussed mainly on substitution at the *meta*-position, and α was defined in terms of partial rate factors by the equation:

$$\alpha = \frac{\ln (f_m^{CH_2X}/f_m^{CH_3})}{\ln (f_m^X)}.$$

Values of α for a number of substituents X were given as follows:

Transmission factors

$-X$ class	$-\overset{+}{N}Me_3$ $-I$	$-\overset{+}{A}sMe_3$ $-I-M$	$-(4-\overset{+}{C_5H_5N})$ $-I+M$	$-\overset{+}{N}H_3$ $-I$	$-\overset{+}{P}Me_3$ $-I-M$
α	0·59	0·56	0·56	0·49	0·44
$-X$ class	$-Cl$ $-I+M$	$-CCl_3$ $-I$	$-CO_2Et$ $-I-M$	$-NO_2$ $-I-M$	
α	0·37 or 0·33	0·28	0·20	0·17	

The assumption was made that substituent effects can be analysed in terms of separate contributions from field and inductive effects, defined by the authors thus: 'By the field effect term (F_X), we mean the change in the free energy of activation produced by the electrostatic interaction between the pole or dipole of the substituent and the charge on the aromatic ring or on the electrophile in the transition state. By the inductive effect term (I_X), we mean the change in the free energy of activation deriving from a modification of the electro-negativity of the 1-carbon atom as a result of the difference in the polarity of the C–X and C–H bonds.' The transmission factor α was then assumed to be the sum of contributions from the two effects:

$$\alpha = x\alpha_F + (1-x)\alpha_I,$$

where α_F and α_I are the transmission factors for the field effect and the inductive effect respectively, and x is the fractional contribution of the field effect to the total substituent effect.

$$x = F_X/(F_X + I_X).$$

Appendix

Since α_I is usually regarded as independent of the substituent, the inconstancy of α (see the values quoted above), particularly as between positively charged and neutral substituents, must arise either from variation of α_F, or of x, or of both quantities.

In order to interpret the results further, the effects of the hypothetical dipolar substituents $-\overset{+\,-}{AB}$ and $-CH_2\overset{+\,-}{AB}$ on the Gibbs' function of activation were calculated on the assumption that each pole of the dipole produces a change in the Gibbs' function of activation equal in magnitude, but in the case of the $\overset{-}{B}$ pole opposite in sign, to that produced by the $-\overset{+}{N}H_3$ group at the appropriate distance. In this way (when allowance was made for conformational effects) a transmission factor of 0·27 was estimated for the group $-\overset{+\,-}{AB}$. This is comparable with the values of α for dipolar groups given above, and suggests that some at least of the variation in the transmission factor between charged and neutral substituents comes from a change in α_F. The inductive effect cannot be ignored for substituents X, because this would, for instance, leave unexplained the greater reactivities produced at the *meta*-position by $-NO_2$ than by $-\overset{+}{N}H_3$. For substituents CH_2X on the other hand, the positively charged substituents are all more deactivating than the dipolar ones; indicating that α_I may be very small.

10.2

Davies and Warren[4] have investigated the nitration of naphthalene, acenaphthene and eight dimethylnaphthalenes in acetic anhydride at 0 °C. Rates relative to naphthalene were determined by the competition method, and the nitro-isomers formed were separated by chromatographic and identified by spectrophotometric means. The results, which are summarised in the table, were discussed in terms of various steric effects, and the applicability of the additivity rule was examined. For the latter purpose use was made of the data of Alcorn and Wells (table 10.2) relating to the nitration of monomethylnaphthalenes at 25 °C. The additivity rule was found to have only limited utility, and it was suggested that the discrepancies might be due in part to the

Relative rates and isomer proportions

Dimethyl-naphthalene	k_{rel}*	Nitro-isomer proportions (%)†						
		1–	2–	3–	4–	5–	6–	8–
1,2–	26·1	—	—	5·0	64·0	13·0	—	18·0
1,3–	13·8	—	8·3	—	81·8	9·9	—	—
1,4–	5·5	—	—	—	—	100	—	—
1,5–	10·3	—	17·2	—	82·8	—	—	—
1,8–	15·1	—	92·7	—	7·3	—	—	—
2,3–	13·6	72·7	—	—	—	21·1	6·2	—
2,6–	15·8	81·6	—	1·9	16·5	—	—	—
2,7–	16·1	75·2	—	3·9	20·9	—	—	—
Acenaphthene	32·2	—	—	35·9	—	64·1	—	—
Naphthalene	1·0	91·9	8·1	—	—	—	—	—

* Estimated relative error ±5 %.   † Estimated error, as expressed, ±2 %.

difference in the temperatures at which the studies were conducted. A more serious problem may be the uncertainty about whether or not the two sets of compounds share a common mechanism, even at the same temperature (§5.3.3).

10.4.2

An attempt to provide a model for theoretical calculations of aromatic reactivity which avoids the extremes of the 'isolated molecule' and 'Wheland intermediate' representations has been reported by Chalvet, Daudel and their co-workers.[15] The attacking reagent was represented by a single atomic orbital, which was included in the linear combination of atomic orbitals from which the molecular orbitals were formed. The calculations were carried out by a 'generalisation of the Hückel method which can be derived in terms of a self-consistent field theory', but full details were not given. The Coulomb integral α_X of the attacking group and the resonance integral β_{rx} between this group and the positions attacked were written in terms of the standard α_0 and β_0 for benzene using the equations:

$$\alpha_x = \alpha_0 + h\beta_0,$$
$$\beta_{rx} = k\beta_0.$$

The value of k was fixed at 0·5 and the π electron energy when the orbital representing the attacking reagent was positioned near to a particular position in the aromatic nucleus was computed, using values of h varying from -3 to $+3$.

Calculations for electrophilic substitution in the quinolinium ion[15b] can be compared with experiment, and for a range of values of h the predicted order of positional reactivities, $5 \sim 8 > 6 > 3 > 7$, agrees moderately well in a qualitative sense with the observed order of $5 \sim 8 > 6 > 7 > 3$ (table 10.3). Further evaluation of the method must await the results of more extensive calculations for a range of aromatic systems.

REFERENCES

1. Personal communication.
2. Cooksey, A. R., Morgan, K. J. & Morrey, D. P. (1970). *Tetrahedron* **26**, 5101.
3. Blackstock, D. J., Fischer, A., Richards, K. E., Vaughan, J. & Wright, G. J. (1970). *Chem. Comm.* p. 641.
4. Davies, A. & Warren, K. D. (1969). *J. chem. Soc. B*, p. 873.
5. Robinson, R. & Thompson, H. W. (1932). *J. chem. Soc.* p. 2015.
6. Robinson, R. (1970). *J. chem. Soc. B*, p. 1289.
7. Thompson, M. J. Unpublished work.
8. Hartshorn, S. R. & Thompson, M. J. Unpublished work.
9. Vaughan, J., Welch, G. J. & Wright, G. J. (1965). *Tetrahedron* **21**, 1665.
10. Hartshorn, S. R., Moodie, R. B. & Schofield, K. In the Press.
11. Banthorpe, D. V. (1970). *Chem. Rev.* **70**, 295.
12. Katritzky, A. R. & Topsom, R. D. (1970). *Angew. Chem. (Int. Edn.)* **9**, 87.
13. Sir Christopher Ingold, *Structure and Mechanism in Organic Chemistry*. 2nd edition, ch. 6. London: Bell.
14. (a) De Sarlo, F. & Ridd, J. H. In the Press.
    (b) Grynkiewicz, G. & Ridd, J. H. In the Press.
    (c) De Sarlo, F., Grynkiewicz, G., Ricci, A. & Ridd, J. H. In the Press.
15. (a) Chalvet, O., Daudel, R. & McKillop, T. F. W. (1970). *Tetrahedron* **26**, 349.
    (b) Chalvet, O., Daudel, R., Schmid, G. H. & Rigaudy, J. (1970). *Tetrahedron* **26**, 365.

Author index

In this Index the numbers of pages on which references appear are italicised. The letter A indicates that the author's work is referred to in the Appendix.

Author index

Author index

Author index

Subject index

Individual compounds or groups of compounds are mentioned in this Index only if some features of their behaviour have received special mention in the text. Individual compounds are listed in the following Compound Index. The letter A indicates that a topic is referred to in the Appendix.

Subject index

Compound index

The letter A indicates that a compound is referred to in the appendix.

Compound index

Compound index